Checklist
for Sustainable
Landscape
Management

Final report of the
EU concerted action AIR3-CT93-1210:
The Landscape and Nature Production Capacity
of Organic /Sustainable Types of Agriculture

J.D. van Mansvelt
M.J. van der Lubbe

Department of Ecological Agriculture
Wageningen Agricultural University
Wageningen, The Netherlands

Granted by
The European Commission,
DG VI
Department of Rural Development

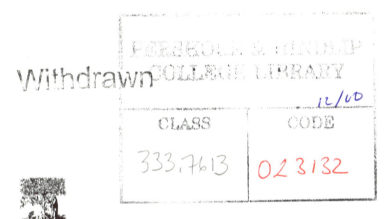
ELSEVIER
Amsterdam-Lausanne-New York-Oxford-Shannon-Singapore-Tokyo

ELSEVIER SCIENCE B.V.
Sara Burgerhartstraat 25
P.O. Box 211, 1000 AE Amsterdam, The Netherlands

First edition 1999
Second impression 1999
Library of Congress Cataloging in Publication Data

Checklist for sustainable landscape management: final report of the EU concerted actio AIR-CT93-1210: the landscape and nature production capacity of organic/sustainable types of agriculture/edited by J.D. van Mansvelt, M.J. van der Lubbe. --1st ed.
          p.   cm.
Includes bibliographical references (p.   ).
ISBN 0-444-50159-2
1. Land use, Rural--Environmental aspects--European Union countries--Planning.
2. Sustainable agriculture--European Union countries--Planning.
3. Landscape protection--Economic aspects--European Union countries--Planning.
4. Sustainable development--European Union countries--Planning.
I. Mansvelt, J.D. van (Jan Diek) II. Lubbe, M.J. van der.
HD588.C47 1999
333.76'13'094--dc21   98-48780
                                                             CIP

ISBN: 0 444 50159 2

♾The paper used in this publication meets the requirements of ANSI/NISO Z39.48-1992 (Permanence of Paper).
Printed in The Netherlands.

CONTENTS

PREFACE

This study was primarily done to find out what would be the overall requirements for a sustainable management of the rural area's landscape. Could they be brought together in a comprehensive system with sufficient consistency to comply with the notion that the landscape is a whole, with character and identity of some sorts and which should be managed accordingly? For this study expertise has been derived from all disciplines and most EU countries. Co-operation of the scientific experts with those from practice and alternating plenary reporting with subgroup visits to farms in the rural landscapes of the participants' countries, allowed for the development of some truly interdisciplinary teamwork. As a second goal of this study, organic agriculture has been included in the theoretical considerations as well as in the observations of practice to find out how organic agriculture contributes to the rural landscape. Including the experience of organic agriculture in the dialogue and looking directly into its contributions to the demanded landscape values, proofed a valuable tool to widen the scope of all those involved.

Starting with the idea that the product of this effort could be condensed into a comprehensive set of general standards for rural land-use, we came to realise more and more that the variety of landscape demands a highly differentiated approach. Moreover, we met the increasing wish of European society to elaborate the regional landscape's different identities' characteristics, which we persistently supported. The solution we found was to re-organise the valuable set of targets, criteria and parameters found as a checklist for sustainable landscape management. The main difference between the two options is that the checklist is meant as a tool for those in charge for any kind of landscape management, to screen all aspects of landscape involved in their development, planning and maintenance. The idea is that, in a multidisciplinary team, together with people from actual and future practice, the checklist should be explicitly elaborated, creating transparency and acceptance of all decisions made on the points mentioned. If the environmental targets, as first mentioned in the checklist, are sufficiently warranted then the freedom to choose implementation of the other targets and criteria should be fairly sufficient to allow to pursue different landscape features, which emphasise in favour of the further development of the landscape's typical identity.

From the discussion in the plenary and subgroup meetings it could be concluded that the multitude of targets, as provided from each of the participating disciplines, would not necessarily be incompatible. By careful localisation, temporisation and scaling, more win-win options emerged as feasible than previously presumed by many. Study of the theory and practice of organic agriculture appeared definitely worthwhile to get inspiration on ways many functions could be served in a viable, sustainable way.

The authors are very grateful for the yearlong co-operation of all participants from science and practice that contributed so much of their time and knowledge into this quite demanding effort. They appreciate the support of the EU in general and its responsible officers in charge in particular for their patient and ongoing support. When this study contributes to the dialogue and landscape management in favour of people and nature of the coming generations, its fully serves its purpose. When it is regarded as fairly normative, setting clear targets that may not yet nor easily be reached, it is fully understood indeed. In order to redirect rural development, agricultural land-use and landscape management away from its definitely unsustainable track of the recent past, norms, targets and policies have to be changed definitely. In as far as nature has the character of a chicken with golden eggs, an eternally productive Sampo (from the Finnish Kalevala) or just the source of life on earth, then caring for nature's well being may serve our successors better then killing it. In as far as humanity is invited to develop nature's potential properties according to its own consciousness development, precisely the challenge to make the chicken eager to produce all golden eggs it needs, complies with the previous statement. Learning from nature and one another to develop all human artistic / creative capacities in favour of our common future is still a major challenge for the next century and on.

During the finalisation of this report we were informed on the existence of the Group of Bruges and their report "Agriculture, un tournant necessaire" (1996). Our study adds up to the studies of the European Council for Nature Conservation (ECNC, 1994) and to the demands of the Council of Europe's European Landscape Convention (1998), which are mentioned in the first chapter of the report. It refers to the multiple objectives of the cluster agriculture – land-use – landscape – rural development. More clearly than the other reports it stresses that no particular technical solution, be it agronomic, environmental, financial or juridical, can be enough to redirect agriculture toward the objectives that society now demands. An integrated, comprehensive approach is needed that merges solidarity, diversity, complexity, connectivity and intention: all issues are included in this report together with the draft of a tool for implementation.

The research strategy for the next decade, as developed by the International Conference of the Dutch Association of Landscape Ecology (WLO, 1998) agrees with the Group of Brugge that changes in land-use, resulting in change of landscape, demands a change of society and its institutions. Finding out about the matching of scales and the integration of interacting functions are the crucial themes in their research recommendations. For the cultural landscape they recommend, among others, the "Development of a handbook of the landscape management activities that enhance and safeguard the valuable cultural landscapes of Europe as an integrated part of sustainable land-use". Moreover, they recommend the "Development of a multifunctional approach in landscape ecology, including socio-economics and disciplines that deal with culture, in order to contribute to new concepts for integrated land-use, including optimisation of ecological functioning of mono-functionally orientated land use". This report offers at least a draft tool for precisely those research objectives, in the context of the demand of the Group of Bruges, the Council of Europe and the EU's revision of its Common Agricultural Policy (CAP). In other words: it offers an interdisciplinary, cross cultural, Europe wide calibrated checklist. One that may serve both as an analytical tool of reference as well as a design tool for local, regional and European policy making on sustainable developments. We hope with this report and especially the checklist to contribute to the dialogue and looks forward for comments that can serve an improved version to appear in the year 2000.

Particular gratitude deems my co-author, Marja van der Lubbe. She, starting early 1998 as an interested outsider to the Concerted Action, helped with the final overall editing and contributed indispensably to clarify many considerations and arguments that had become implicit within the team over the years of the Concerted Action's co-operation. Also Darko Znaor and Juliete Kuiper, who each from their major science realms' backgrounds helped to bring all aspects together, deserve a very special "thank you" for their input to the Concerted Action. Darko Znaor mainly contributed to the writing about the environmental and ecological criteria of sustainable landscape management (section 3.2.1) and to the composition of Annex 2 ("Compliance of the checklist of sustainable landscape management with some other standards for organic/sustainable agriculture"). Juliette Kuiper contributed a lot to the writing of section 3.2.3 (Criteria for the humanity realm: psychology and physiognomy/cultural geography).

Finally it should be stressed that without the dedicated participation of all international experts mentioned in the annex, and the financial support of the EU's commission, this work would have never succeeded.

Who ever benefits form the existence of this report should be aware of the gratitude they all deserve.

Jan Diek van Mansvelt

September, 1998

SUMMARY

In the mid-nineties several EU and national institutions published reports demanding full attention for the deteriorating and vanishing European landscapes. Therein, not only industrialisation, urbanisation and increased infra-structural density were at stake but more than before, attention was drawn to the role of agriculture in development and management of the landscape in the rural areas. It was discovered that farmers are neither paid nor educated for landscape production, although they actually manage most of the landscape's surfaces, neither are most of their organisations and institutions. However, it was also discovered that for the population of a country, urban and countryside alike, the rural landscape was important for tourism, sports, recreation, identity development and quality of life. These functions of the landscape were discovered as additional functions, besides the well know commodity production (food and fibre) and the related functions. As relatively new, a set of environmental and ecological functions was described, such as the production of drinking water, rainwater retention, CO_2 fixation and on site preservation of bio-diversity.

From all those reports, the need for a new policy favouring the multi-functionality of the landscape emerged. The subsequent demand for redirection of land-use went along with proposals for payment schemes to give farmers incentives for well targeted landscape management. Moreover, such incentives could provide them with income support in a period where food-prices went down because of the opening of the EU market for World Trade. In that situation, a Concerted Action plan was offered to the EU in the framework of their AIR3 Program to bring together an international and interdisciplinary team of experts involved in landscape research and to elaborate a scheme for landscape value assessment that could provide keys for landscape payments. The plan was made by the Department of Ecological Agriculture (DEA) of the Wageningen Agricultural University (WAU), accepted by the EU and started late in 1993 with a first plenary meeting of 25 participants from 9 countries. The Concerted Action was finished with a presentation of the results at the International Conference "Things to do: proactive thoughts for the 21st century" of the Dutch association for Landscape Ecology (WLO) organised in Amsterdam 1997. This report summarises the total work done by the Concerted Action *The landscape and nature production capacity of organic/sustainable types of agriculture*.

The objectives of the Concerted Action were:

1. To bring together a group of experts from different European countries to identify and validate agro-landscape and a compliant interdisciplinary set of features that are important for the assessment of sustainable agro-landscape planning and management in the EU.
2. To establish an interdisciplinary list of criteria for the assessment of sustainable landscape management in the EU.

3. To indicate organic agriculture's potential contribution to the sustainable landscape management in the EU.
4. To discuss the established checklist and the assessed performances with experts from different research disciplines, planning and practice.
5. To disseminate the results in oral presentations and (refereed) publications.

The objectives are addressed using an unifying concept derived from Maslow's study on human motivation translated to the landscape and perceived as a reflection of the priorities and motivations leading the actions of the people that were de facto in charge of the landscape during the previous decades. This unifying concept is used as a frame of reference to integrate various disciplines, from the natural, social and humanist science realms, as well as various levels of action and knowledge, from academia to farmers' practice. By always relating parameters with criteria and criteria with targets, specifying the relevant scales in time and space, some comprehensive contextualisation can be achieved. Thereto, iterative alternation of plenary meetings in Wageningen and subgroup meetings in member countries, with visits of some farms in their landscapes, proved to be an excellent tool to bring disciplinary, abstraction level and national perceptions together.

To identify and validate agro-landscape through an interdisciplinary set of features that are important for the assessment of the sustainable landscape planning and management in the EU, an international team of landscape oriented experts from EU-member states, which represent different disciplines, has been formed. By mutual presentation of their (disciplines') perceptions of landscape functions and values, the required qualities and parameters for assessment, common denominators and differences are identified. Mutually questioning and discussing the points of view, positions and value systems underlying the differences, created much transparency and solved many of them. The considerable added value of understanding one another's points of view has been appreciated.

As major realms of a scientific approach natural (β), social (γ) and humanist (α) sciences are included. In each of those two subsections are discerned:
- Environment and ecology (biosphere) in the β realm, referring to the resource conditions and biological relationships;
- Economy and sociology in the γ realm, referring to the flows of finances and services and the participative procedures;
- Psychology and physiognomy with cultural geography in the α realm, referring to subjective regional landscape appreciation and objective regional landscape identity.

Further, a checklist with six columns, which are in need of assessment, has been defined with in each column the major criteria for sustainable landscape management.

A summary of the Table of the Checklist is presented below.

Summary of the Table of the Checklist presented in chapter 3

Criteria for the development of sustainable rural landscape management					
Quality of the (a)-biotic environment		*Quality of the social environment*		*Quality of the cultural environment*	
Environment	Ecology	Economy	Sociology	Psychology	Physiognomy
1.1 Clean environment 1.2 Food and fibre sufficiency and quality 1.3 Regional carrying capacity 1.4 Economic and efficient use of resources 1.5 Sustainable, site-adapted and regionally specific production systems	2.1 Bio-diversity 2.2 Ecological coherence 2.3 Eco-regulation 2.4 Animal welfare	3.1 Good farming should pay-off 3.2 Greening the economy 3.3 Regional autonomy	4.1 Well-being in the area 4.2 Perma-nent education of farmers 4.3 Access to participatio n 4.4 Access-abillity of the landscape	5.1 Compliance to the natural environ-ment 5.2 Good use of the landscape's potential utility 5.3 Presence of naturalness 5.4 A rich and fair offer of sensory qualities 5.5 Experiences of unity 5.6 Experienced historicity 5.7 Presence of cyclical develop-ments 5.8 Careful management of the landscape	6.1 Diversity of landscape components 6.2 Coherence among landscape elements 6.3 Continuity of land-use and spatial arrange-ment

Within each of these columns a number of targets have been defined per landscape aspect, from which criteria have been derived. Where possible appropriate parameters have been indicated to assess the criteria. All in all, some 200 criteria have been identified and discussed. To allow and warrant appropriate use of those criteria, that is a use according to the Concerted Action's team of experts, each of them has been discussed in some detail to indicate its conceptual context including its scale in time and space. Thus, a long checklist with explanations has been established. From that full checklist a dried-up version has been developed showing only the targets, criteria and parameters, to allow a type of scoring as to be chosen by the user.

Subsequent versions of the Table and the Checklist have been established through meetings of subgroups visiting some organic and non-organic farms in their landscapes. In walks over and around those farms and in discussions with the people in charge of the farm management, the checklist was over and again assessed in its feasibility for quick scans and in depth comparison. Including local experts of agriculture, environment, ecology, landscape etc. in the farm visits and quick scan discussions allowed for a check on the feasibility of the checklist for each of the visited particular European regions. This iterative process of alternating plenary team sessions with local expert meetings, the earlier target of going for a set of landscape management standards has been transformed into the adapted final target of going for a checklist. The difference is that the checklist presumes to be applied and used by local stakeholders and experts from all relevant disciplines, thus relying on local knowledge for the fine tuning of the management within the transparent value system offered by the checklist. As the checklist presented in this report covers such a wide range of comprehensive values, any top-down enforcement would threaten people's fair demand for local empowerment and self-responsible decision-making.

As an interesting spin-off of making and assessing the checklist in practice, the Concerted Action team has found evidence that the targets and criteria, as derived from all participating disciplines, were not necessarily incompatible. There are indications that a certain degree of synergy can be established between all of the targets, which will emerge farm and local management to go for sustainable multifunctionality. In depth analyses and further confirmation in practice is needed to find out about the extend and limits of this synergy.

Based on the comparisons of organic agriculture's landscape performance, although by and large limited to quick scans by international and local expert teams, the Concerted Action's team concluded that in theory and practice organic farming can importantly contribute to the sustainable management of appreciated landscape values. As the standards for organic agriculture do not yet include much specified standards for the landscape as a product of agriculture, the attitude of the farmer is still most decisive in effectively using all the landscape opportunities of that approach to agriculture.

As an additional product of the Concerted Action, an overview of possible uses and users of the checklist is presented in the fifth chapter of the report. As the redirection of agriculture in Europe is in no way a question that can be solely solved by changing some techniques, the use of the checklist, with its elaborated and explained validation system, can very well serve to raise the awareness about the interrelations of landscape functions and features. The wide range of possible uses and users indicates the wide range of stakeholders involved in the landscape use and its management. By offering at least an outline of a unifying concept, the use of this checklist can decrease conflicts and incompatibility of uncoordinated and counter-productive actions.

Systematic evaluations of the use of the checklist will allow for its improvement over the next years, for which the authors would be pleased to serve. Funding for experiments in the use of the checklist for land-use planning and landscape management, now with special attention to the regional identities and qualities, would importantly enhance the possibilities for its improvement through systematic evaluations as mentioned.

The results of the Concerted Action are disseminated through the following sources:

- Proceedings of the plenary meetings
- Journals,
- A publication titled *Checklist for sustainable landscape management* (Elsevier),
- Dialogues and discussions with farmers and regional experts,
- Dialogues and discussions with colleagues at conferences.

In November 1998, the framework and checklist will be presented at the 12th International Scientific IFOAM Conference, titled *Credibility of organic agriculture in the 21st century*, and organised in Mar del Plata, Argentine.

CHAPTER 1 INTRODUCTION

1.1 BACKGROUND AND PROBLEM STATEMENT

In view of the recent discussions in the EU on the need for redirection of agriculture toward extensification and sustainability, increasingly the question is raised whether beside food and fibre, the cluster environment/nature/landscape should or should not be regarded as a valuable product of agriculture. Quite often, the arguments in this discussion are summarised as the segregationist (separation of land-use functions) versus the intergrationist (merging land-use functions). Within the agricultural sector, which, from the 50ties onward has been rather food-production oriented the contribution of agriculture to landscape values must be included now as a valuable asset in its multi-functionality. That is: over-all agro-landscape studies in addition to studies on agriculture performance in environmental issues and its production of 'elements of nature' (Van Mansvelt and Van Laar, 1998).

In 1993, the Institute for European Environmental Policy, published a report for the Dutch Ministry of Agriculture, Nature Management and Fisheries, entitled Nature Conservation and New Directions in the EC Common Agricultural Policy and written by Baldock and Beaufoy. Their report indicates the general context of the problem statement of the EU concerted action AIR3-CT93-1210 *The Landscape and Nature production Capacity of Organic/Sustainable Types of Agriculture*. Baldock and Beaufoy stated in their report that the rationalised intensive agriculture has often been associated with damage and destruction of the environment, natural and semi-natural habitats and (visual) landscapes. Whereas on the other hand Europe's valuable habitat and landscape diversity is a product of Europe's long time agricultural management in the previous centuries. From this statement it is derived that agriculture should be addressed according to an integrated approach. Thereby it is not only important to prevent the existing and prevent upcoming negative effects caused by agriculture, such as pollution, habitat degradation, noise, smell and soil erosion. But even more to maintain such positive functions of nature and landscape produced by agriculture, as there are the historic creation and management of semi-natural habitats and landscapes of high environmental and amenity value over large areas of the European Community. Baldock and Beaufoy (1993) stressed that maintaining the positive functions of nature and landscape requires more than just financial support to farmers. In fact, it requires steering agricultural development in such a way that farming systems as well as individual farm practices, that is forestry and land-use in general, of each region in the European Union (EU), be in harmony with the integrated nature conservation objectives. This means that integration should not be confined to minimise the negative environmental impacts of agriculture, but also that it should aim to maintain and increase, as far as possible, agriculture's positive nature conservation and landscape functions. People who know about the production standards of organic agriculture, set by IFOAM and the EC-regulation, will recognise and admit that organic agriculture is an interesting basis for the required integrated approach for agriculture. This, because organic agriculture goes for more of the better instead of going for less of the bad (Van Mansvelt and Mulder, 1993). Realisation of such a new and integrated direction in the Common Agricultural Policy (CAP) of the EU requires a compatible research strategy, that warrants that all parties involved in the new policy work together, instead of competing in a contra-productive way for the best solutions according to their own disciplines. Thus, a scientific approach is needed that integrates all different aspects in a consistent, communicative and convincing way.

Baldock and Beaufoy (1993) mentioned also that, in order to design an appropriate strategy that meets the mentioned objectives of banishing the negative environmental effects and maintaining its positive functions, nature and landscape values of various farming systems throughout the EU should be assessed. The limited perspectives of a largely "N-driven" agriculture are based on the application of high external inputs like fertilisers and pesticides, extremely high stocking rates and the removal of all areas that are monetary non-productive. The rationality underlying this agricultural policy increasingly tends to off-soil production of crops and animals with fully computer controlled environmental conditions. Organic agriculture counteracts and avoids these problems of modern agriculture, by going for co-operation with nature, using its ecosystem's regulating strategies (Lampkin, 1990; Van Mansvelt and Mulder 1993). From this way of thinking about agriculture, through thorough discussions with practitioners from many countries world-wide, a system of clear and consistent standards has been derived, as reflected in the IFOAM standards (IFOAM, 1996) and drawn upon in the EC Regulation 2092/91.

Baldock and Beaufoy (1993) listed nature and conservation implications of various intensive and extensive practices of agriculture. They show that considerable similarity can be found between the practices of farmers licensed as organic and those practices of extensive agriculture that are recommended for their positive effect on nature and landscape. However, so far little research has been done about the actual compliance of organic agriculture and the need for nature and landscape management.

Incentive payments are the predominant mechanism currently used in the EU for the promotion of environmentally sound farming. The new EC Regulation 2078/92 gives considerable additional momentum to the incentive payment approach in the EU. At the moment, agro-environment incentive payments can have various objectives, such as the maintenance of particular habitats or landscape types or the reduction of agricultural pollution (excessive nutrients or pesticides). This can easily lead to incompatible sets of payments, with contra-productive effects, depending on the interests involved in the implementation of the EU 's regulations in its member states. On the other hand, different objectives such as environmental, landscape and habitat protection may not be clearly distinguished, also leading to a decreased effect of the payment schemes.

Some of the objectives of agro-environment incentive payment schemes are summarised as follows by Baldock and Beaufoy (1993):

- Extensification of production practices for a mixture of broad objectives such as reducing agricultural production and lessening the negative environmental impacts of farming, for example overgrazing, high use of chemical inputs, etc.
- Maintenance of existing landscapes and habitats.
- Environmental enhancement of an area.
- Management of abandoned farmland and woodland.

The different options for incentive payments refer to the EC Regulation 2078/92. They propose extensification of few- or single-commodity production systems, in favour of a multi-objective approach and the maintenance of the traditional farming systems that harmonise with nature, and in favour of the re-introduction of valuable landscape management practices in abandoned areas. However, these incentives focus on single objective payments as presently developed in different member states, instead of elaborating the integrated approach as they suggested in other parts of their report. Nevertheless they mentioned organic agriculture as an interesting option for such an integrated approach. Although a license for organic farming does not necessarily result in high nature conservation benefits, as the current standards of organic agriculture are indeed very modest in specifying standards for nature and landscape, the farmers applying for recognition as organic farmer will usually go for it. Thus the organic label tends to provide a sound basis for high natural value (HNV) farming (Van Mansvelt and Mulder, 1993; Van Mansvelt and Stobbelaar, 1997) See also section 4.2 ("Empirical data collected from literature").

While Baldock and Beaufoy (1993) give a general context of the problem, ECNC (1994), a network project of the European Centre for Nature Conservation (ECNC) entitled "Natural environment and sustainable development: habitats, species and human society", presents the political context of the problem. There the ECNC stated that: "Since agriculture traditionally depends on sound environmental conditions, farmers have a special interest in the maintenance of natural resources. For centuries they maintained a mosaic of landscapes which protected and enriched the natural environment" (ECNC, 1994). Clearing and levelling of the land, focusing on mono-cultures, together with intensive use of fertilisers and pesticides, have resulted in losses of landscape, habitat and species diversity. A degradation of landscape diversity into rather monotonous and uniform areas of intensive agriculture and on the other hand a-specific wildernesses on abandoned land, are the results of recent policy on the European landscape. Although the CAP included the agri-environment in its 1992 reform, the processes that had been started through previous policies are neither easily re-directed, nor are its damages easily made undone. Moreover, several weaknesses in the phrasing and implementation of the extensification schemes lead for example to losses of diversity in uncontrolled set-asides and single-species refore-station at former agricultural lands (ECNC, 1994). These effects damaged the social, historical and cultural diversity of many regions that until then had been one of the many strengths of the European Community, providing the citizens with a deep sense of identity. The European community has consistently recognised the need for special efforts to help the less developed regions and to encourage these regions to help themselves. Emphasising regional autonomy, with a relative self-sufficiency in food and fibre per region, is an important theme in this policy. Low external input sustainable agriculture (LEISA), based on the ecological principles of organic agriculture, could thereby used to play a crucial role.

ECNC (1994) wonders if "Natura 2000", the EU strategic plan for nature and landscape diversity throughout pan-Europe, is enough to meet the global requirements for rural development and if not what the alternatives might be. In their opinion, "Natura 2000" is in danger to focus too much on hotspots of natural quality, leaving the wider landscape open to degradation. Another aspect, raised by ECNC, is how to link requirements of the natural environment to the social requirements of employment, tourism, education and training. Here again, organic agriculture, as multifunctional type of agriculture, could be used as a good example for possibilities to link the requirements of the natural environment to such social requirements as employment, tourism, and education. Also the United Nations Economic Conference for Europe (UNECE) stated in 1995 that, recognising the uniqueness of landscapes, ecosystems and species, which include economic, cultural and inherent values, a pan-European approach to the conservation and sustainable use of shared natural resources should be applied. The UNECE invited the Council of Europe and the UNEP, in co-operation with OECD and IUCN, to establish a task force or other appropriate mechanism in order to guide and co-ordinate the implementation and further development of such a strategy.

Also the Pan-European Biological and Landscape Diversity Strategy (PEBLDS, 1995) indicates the political context of the problem of the Concerted Action. One of the challenges to be addressed, according to the PEBLDS (1995), is to prevent further deterioration of the landscapes and their associated cultural and geological heritage in Europe and to preserve their beauty and identity. This challenge goes together with correcting the lack of integrated perception of landscapes as a unique mosaic of cultural, natural and geological features, and with the establishment of a better public and policy-maker awareness together with a more suitable protection status of the landscape and nature features throughout Europe. The following aspects are mentioned by the PEBLDS (1995) to realise the opportunities for the cultural and social commitment to maintain local and regional individuality, as expressed by cultural and geological heritage features in the landscape:

- Compile a comprehensive reference guide on European biological and landscape diversity, to further develop and seek acceptance of criteria to identify priorities for conserving geological and cultural landscape features. Identify traditional agricultural and related landscape management types and assess the effects of marginaliation or intensification of the landscape (1996-1997).

- Establish guidelines, following assessments and evaluation, to address policies, programmes and legislation for the protection of cultural heritage, geological heritage and biological diversity that are mutually supportive and complementary, and use them to their full potential in the conservation of the landscape (1996-2000).
- Set up a code of practice to involve private and public landowners to promote awareness of the relevance for bio-diversity of landscapes traditionally valued and managed for their historical and cultural importance, focusing on historic parkland estates and historic buildings (1996-1998).
- Establish an action plan using awareness techniques, guidelines and demonstration models to safeguard geological features in the landscape, actively involve and consult landowners and the energy, industry and water management sectors in their conservation (1996-1998).
- Investigate the relationship between traditional landscape and regional economy. Develop a framework to stimulate initiatives for regional development based on landscape diversity, involving eco-tourism and traditional crafts. Find successful case studies and set up programmes for exchange of expertise (1996-2000).

Also a study of De Putter (1995), entitled *The Greening of Europe's Agricultural Policy*, subscribes the political context of the problem, by mentioning that under the regulation 2078/92 EU-member states have decided upon a structural, premium supported policy in favour of extensification of agricultural practices, environmental protection and nature and landscape preservation. Thereby the member states agree that farmers are no longer food-producers only, but also caretakers of the environment, countryside and landscape. Possible income losses generated by the additional farmer tasks will be compensated by yearly premiums to farmers. Apart from benefits for the environment at large, these regulations are also mentioned to contribute to balance the surplus market of agricultural products. Altogether, the multiple goals of the 2078/92 regulation demand a holistic and consistently integrated approach. However, as such an approach supposes a complicated monitoring system, the commission focuses on (semi-)single target programmes. De Putter (1995) mentions that organic agriculture is a feasible instrument to reduce negative environmental impacts, for instance by reducing external chemical inputs, and that it contributes to balance the market of agricultural products. However, she does not mention specifically that organic farming may and actually does contribute positively to landscape preservation and to the prevention of agricultural decline and natural hazards. Other activities, like environmental practices, maintenance of countryside and landscape, upkeep of abandoned farmland or woodland and set aside (twenty years), are regarded, by De Putter (1995), to cover landscape preservation and to the prevention of agricultural decline and natural hazards. Although many of such practices are actually applied in practice by organic farmers, they are hardly implemented as such in organic production standards indeed. Nevertheless, several authors recognise that organic farmers are positively active in these fields (Beissmann, 1997; Van Mansvelt and Mulder, 1993).

A more scientific context of the problem statement is, for instance, given by Bockemühl (1992), Colquhoun (1997), Giorgis (1995), Naess (1989), Seel (1991),

Stroeken *et al.* (1993), Vahle (1993) and Vos and Stortelder (1992). Holistic approaches elaborated by the above mentioned researchers have been consulted for an integrated approach of the multifunctional landscape validation and assessment. From that consultation it seems that an interdisciplinary approach to meet the requirements of a sustainable landscape management, should include contributions from natural, social and humanistic sciences. Especially the work done by Giorgis (1995) shares our opinion about interdisciplinarity, in an approach that largely complies with the approach developed by the Concerted Action at stake. Giorgis (1995) mentioned that in addition to the objective approach of the ecologist or historian, sensorial perceptions can tell a lot about the landscape. People's appreciation is largely subjective and involves value judgements. The future of European landscapes calls for a wide-ranging debate on ecological, economic and cultural values. The countryside has been created, tended and used as living space by the rural community in addition to the production of food and fibres and these days it also provides city dwellers with recreation and services (Giorgis 1995).

Moreover Giorgis (1995) mentioned four fundamental principles to be abided by in favour of landscape quality. Within those principles there are several aspects that comply with our Concerted Action, viz.:

1. Respect for life and preservation of landscape diversity
    - Natural resources should be managed in such a way that it enables present and future generations alike to make a living. Advocacy of sustainable development is therefore essential.
    - Maintaining, recreating and increasing the landscape structures, vegetations or mineral structures providing protection against erosion by wind or water.
    - Encouraging farm practices that use and nurture the living properties of the soil: rotational cropping, a balanced use of organic manure, reasonable mineral inputs and recycling of healthy organic wastes by composting.
    - Promoting balanced utilisation of all productive land to prevent the soil being degraded by over-intensive use or dereliction.

Especially for water quality, the following aspects have to be considered and respected:

- Resource management: water is limited in supply. This means that water should be used economically and managed in an efficient and ecologically sound way. Thus an integrated approach is needed.
- Maintaining water quality: pure water is crucial for a healthy environment. With good knowledge of ecological processes, it is possible to preserve water quality without having to apply chemical treatments.
- Encouraging farm practices using little water and causing little or no pollution and limited soil erosion.
- Developing landscape structures, which help the water self-purification process by restoring and consolidating riverside vegetation and developing wooded strips, grass verges and hedgerows.

- Preserving the visible presence of water instead of converting it into concrete pipes or covering over watercourses and ditches.
- Protecting and restoring natural habitats associated with the presence of water and avoiding the use of non-living materials on stream banks.
- Managing in a sustainable way the recreational and educational potential of water in the countryside.

2. Preservation of biological diversity
   - Creating ecologically stable landscape structures at two levels:
     - International and national level: natural areas of major biological interest safeguarded by national parks, reserves and various protected areas.
     - Regional and local level: ensuring continuity and liaisons and providing migration corridors between the different natural features of the landscape like copses, hades, waterside vegetation etc.
   - Supporting countryside stewardship, by making remunerating of the farmers' labour a condition for the signing of management agreements. In particular such agreements that regard areas and habitats which are totally dependent on such type of management. For example, dry grasslands, wetlands etc.
   - Recommending integrated pest control procedures, using natural landscape structures as the habitat of the auxiliary fauna and minimise chemical treatments at crop and timber production plots.
   - Developing institutions for the conservation of genetic varieties, promoting and rearing of rustic breeds, and demonstrating the potential economic values of this diversity.

3. Development of solidarity
   - Demands for space, quality of life and food supplies must be taken into account and shared among all sections of the community. Therefore, productive landscapes accessible for all are necessary.
   - Access to water. Everybody should have access to water for enjoyment as long as it is consistent with the principle of wise use.
   - Ensuring that the economic and ecological potential within each region are balanced in a such a way that it takes the need for better distribution of wealth and of the environmental carrying capacity of the region into account.
   - Supporting locally grown products.
   - Supporting certification of quality and guarantees of origin, which associate with specific agricultural practices keeping landscape quality and preserving the environment.

For the accessibility to all, the following aspects have to be considered:
- Maintaining, protecting and developing networks of ecologically acceptable roads and footpaths in the countryside.
- Promoting highway embellishment policies.
- Introducing landscape awareness into the design and construction of all infrastructures.

For the observance of democratic procedures, the following aspects have to be considered:

- The landscape is everybody's concern. The standard of partnership, negotiating capacity and interchange among the social groups, creating, managing and using the landscape will be shown in the landscape quality.
- Global and integrated water management presupposes a concerted approach by everybody affecting landscape with his or her activities, for example local authorities, farmers, angling clubs, conservation organisations, environmental groups, local inhabitants etc. They should together develop a landscape plan.
- Developing landscape plans, contracts and charters.
- Changeling skills and sensitivities into landscape planning.
- Developing training in landscape design and making technicians sensitive to aesthetics and artistic considerations.
- Fostering community involvement.

4. Respect for regional identity and the right to enjoy beauty
    - Every society is entitled to express its character, genius and ideas about beauty. Landscape is one of the media, which can be used to express these aspects.
    - Each landscape has a visage with specific features, which make the landscape unique in colour, plant life and minerals. Proportions and distributions of empty and non-empty spaces in the landscape have their particular feelings and atmosphere, situated in time and space and for different life-spans. It is important to understand the perceptions of musicians, painters, writers and landscape artists about the landscape.

Specifications on how to implement these requirements and how to assess the implementations have as yet to be established. In particular the translation to the level of farming practices, for single farms or farming co-operatives, should be elaborated to make them compatible with the landscape requirements as phrased. Apart from landscape researchers working toward a better and more sustainable landscape management, there are also agronomists, who extend their perception of agricultural production in order to include the nature production values of agriculture. For example, Vereijken (1994, 1995, 1996a) has been involved in an EU Concerted Action and designed farming prototypes, which comply with six general and twenty specific social values and interests, as perceived by the researchers and farmers involved in the Concerted Action. The main objectives of the integrated agriculture farming system (IAFS), as mentioned by Vereijken (1994, 1995, 1996a), are basic income/profit, environment and food supply. The representatives of the ecological agricultural farming system (EAFS) rank these objectives as follows: 1) environment, 2) food supply and 3 nature/landscape. Vereijken (1994, 1995, 1996a) concluded that IAFS focus on short time strategies whereas EAFS goes for long time strategies. Agro-ecological criteria for I/EAFS farm layout with clear effects on the landscape are presented by Vereijken (1996b) and show that the agro-ecological unity/identity of farms, together with minimising ecological stress (pest-prevention / infrastructure of pest-predator bio-topes) are leading design motivations.

This approach to agriculture is quite comprehensive for the farming practices, yet the integration of those sustainable farming practices in the land-use and landscape management planning has still to be elaborated.

1.2 OBJECTIVES

In the EU's common agricultural policy (CAP), the objectives have been extended to include the landscape production that inevitably goes along with agricultural production. Thus agriculture has become a multitude of objectives as now the environment-, nature- and landscape-production join the production of food and fibre. Landscape production, like any other production, requires appropriate quality standards, which in view of UNCED's global conference (1992), should meet the requirements of sustainable development.

Criteria for the different types and dimensions of landscape quality are found to be scarce and incomplete in literature, especially in as far as the contribution of agricultural land-use to the landscape is concerned. Research about recommendations for landscape and nature production by farmers and how to include these recommendations into standards and payment schemes is strongly needed, but missing so far. Such research, which warrant the necessary integrated multi-objective approach, requires multidisciplinary research covering all relevant aspects of the rural, agro-sylvi-pastural land-use in the landscape and should include organic types of agriculture into its theory and analyses. The incompleteness of research, so far done, is recognised as a constraint for the development and implementation of a cross-compliant strategy in the supporting agricultural policy. Also the lack of research on the compliance of organic agriculture with the demands for landscape and nature management has to be tackled. In view of the actual performances of organic agriculture, this type of agriculture could be a farming system that fits to cross-compliant rewarding. A cross-compliant approach will better allow for a systems' and goal oriented payment schemes, instead of for single solution oriented payment schemes, based on single, prescribed or solely allowed technologies (Bosshard *et al.*, 1997; Van Mansvelt and Van Elzakker, 1994; Van Mansvelt and Mulder, 1993).

In 1993, the EU Concerted Action *Landscape and nature production capacity of organic/sustainable types of agriculture* (AIR3-CT93-1210) started, aimed at identifying and validating agro-landscape features that are important for the assessment of planning and management. Agro-landscape standards, among others, should help to keep the quality of rural areas on an acceptable level (prevention), or even in due course, to restore them to the desired quality level.

A number of issues that need to be examined in more detail emerge from the foregoing discussion:

1. The composition of a list of criteria to be used for the assessment of all kinds of planned or ongoing land-use activities affecting landscape. Agriculture is explicitly included as a land-use activity affecting the European landscape.
2. The contribution of organic agriculture towards landscape quality.

The Concerted Action has been initiated to discuss standards for the assessment of sustainable agro-landscape values in the EU and how to define criteria and parameters for the development of such standards. The insights gained by this Concerted Action programme may help farmers, policy makers, government and politicians to manage the development of agro- and forestry-landscape toward sustainability and socio-cultural appreciation. The general set of standards and the list of criteria and parameters, including regional specifications and examples, can be used as a guide for agro-landscape production and as a framework for the follow-up and updating of those standards, criteria and parameters

Within the overall objectives of the concerted action, separate goals can be distinguished, which have been tackled during the four years of the Concerted Action:

- To identify and validate agro-landscape and a compliant interdisciplinary set of features that are important for the assessment of the sustainability of landscape planning and management in the EU.
- To establish an interdisciplinary list of criteria for the assessment of sustainable landscape management in the EU.
- To discuss the established checklist and the assessed performances with experts from different research disciplines, planning experts, and experts of practise.
- To disseminate the results in oral presentations and publications.

1.3 APPROACH

The EU Concerted Action programme entitled *Landscape and nature production capacity of organic/sustainable types of agriculture* (AIR3-CT93-1210) started in 1993 and continued for four years. Some twenty scientists from 11 EU and neighbouring countries have met in several subgroups in nine different countries to discuss their targets and to compose a set of criteria and parameters, as well as the way in which these can be merged in one structure.

To achieve the above mentioned objectives, the following aspects are taken into consideration:

- An international group of twenty European scientists has been invited to contribute their national and international expertise in order to compose a comprehensive and consistent list of criteria, which fit to the whole European Union.

- A comprehensive multidisciplinary attitude is expected and required from the international group of experts in order to warrant that the developed criteria will fit and be useful in the rural, agro-sylvi-pastural landscape, which they are meant to serve.
- The expertise of the international and multidisciplinary research team should include sufficient knowledge about the basic concepts, the implementation, and the actual performances of organic agriculture in order to warrant the relevance of criteria for further development of organic agriculture.
- Bring the various opinions, positions and experiences of all participants together and discuss the compatibility or incompatibility of their perceptions. From each partner of the expert team, ability and willingness to get involved in an open-ended discussion with experts from other countries and other disciplines, with other points of view, philosophies and attitudes were required, allowing for relevant co-operation toward the objectives of the Concerted Action

An iterative alternation between plenary and subgroup meetings of the international and interdisciplinary team has been used as a method to reach the research objectives, taking into consideration the above mentioned aspects. During the yearly plenary meetings, organised in Wageningen, participants present their landscape perceptions and experiences, which are discussed in view of society's demand for a shift toward sustainable landscape management. These perceptions and experiences should be specified into the important landscape functions and the landscape qualities, necessary to provide the landscape functions. Relevant scales, necessary for the landscape functions, and the parameters used to assess the landscape qualities and the desirable ranges of those parameters should also be specified by every participant. Participants used examples from their own research to present the required information. A preliminary list of criteria and parameters has been composed from these presentations and the discussions afterwards. Gradually, the list of criteria and parameters has been upgraded over the subsequent meetings.

Every year, two or three subgroup meetings were organised by one of the EU participants in co-operation with the other participants of the Concerted Action. During the subgroup meetings, quick scans on some farms were performed in the landscapes of the participating countries, to check on the spot the feasibility and (in)completeness of the list of criteria. Organic farms were purposely included in the quick scan performances. The subgroup meetings were dedicated to common and team-wise agro-landscape observations. The on-site discussions on the criteria and parameters during the subgroup meetings facilitate mutual understanding among the participants representing various opinions, disciplines and countries. Observing, working and discussing together on a common interest object, is a perfect tool to overcome misunderstanding and disciplinary biases. It also balances lob-sided points of view of the various participants. The discussion rounds during the subgroup meetings were done in three sections, viz. one for the natural science oriented aspects, one for the socio-economic science oriented aspects, and one for the human science oriented aspects. Subsequently, in a plenary meeting the three different sections report their findings and discuss the compatibility or incompatibility of their criteria and parameters for the agro-landscape management. Several papers have been produced, based on the subgroup meetings, discussing the landscape values of various farming systems observed during the quick-scan approach and elaborated afterwards. See for instance, the proceedings of the four plenary meetings of the Concerted Action done in 1994, 1995, 1996, and 1997 and the special issue of *Agriculture, Ecosystems & Environment* (1997). Another special issue of *Agriculture, Ecosystems & Environment* is in preparation.

In 1995, the list of criteria and parameters reached a certain point at which the participant of the Concerted Action felt the need to discuss its feasibility with local experts from outside the Concerted Action team, to receive second opinions. With permission of the EU-officers, local and regional farmers, extension workers, policy makers and nature conservation experts were invited and participated in the Concerted Action meetings of last two years. Their observations of the agro-landscape and their opinion about the criteria and parameters have been taken into consideration as well. At the meetings with local experts, the issue regarding the type of users and the kind of uses is envisaged and has resulted in additional workshops and a special chapter in the final report of the Concerted Action.

The list of criteria and parameters produced by the Concerted Action, the so-called "checklist", is extensively compared with existing sets of standards for sustainable agricultural land-use, such as the basic standards for organic agriculture of IFOAM (1992) and the European Union (Regulation 2092/91. Also the criteria of I/EAFS (Vereijken, 1994, 1995, 1996a; Kabourakis, 1996) for sustainable farming are included in the comparison.

1.4   STRUCTURE OF THE REPORT

After the introduction, the research methods are described in chapter 2. The study is based on Maslov's theory about human motivations and on an interdisciplinary approach concerning environment, ecology, economics, sociology, psychology and humanities. This theory and its compliant interdisciplinary approach will reappear in chapter 3, which presents the results of this study, viz.: a checklist with criteria for the ecological realm, the social realm and the humanities' realm of sustainable landscape management. To improve the collection of targets, criteria and data of sustainable landscape management, an overview of the various criteria has been developed. This overview or table with criteria for sustainable landscape management is presented in chapter 3 as the Table of the Checklist (Table 3.1). It is produced in an iterative process of data collection from all kinds of landscape observations and based on Maslow's comprehensive theory of human motivations, adapted to landscape management (Maslow, 1968). Compliant to Maslow's triangle, the checklist covers all three major realms of science: the natural sciences (β), the social sciences (γ) and the humanities (α). These three realms are discussed in detail in section 3.2.1 the natural sciences (β), section 3.2.2 the social sciences (γ) and section 3.2.3 the humanities (α) of chapter 3. The Table of the Checklist, representing different disciplines, is subdivided into six columns, each representing a specific discipline: environment, ecology, economics, sociology, psychology, physiognomy and cultural geography. For each discipline (column), the main targets and criteria are mentioned.

Of course all targets, criteria and parameters of one discipline (column) in the Table of the Checklist (Table 3.1) are often related to aspects and dimensions from the other disciplines (columns). The different realms are integrated into each other and figure in paradigms, which differ between interest groups active in and responsible for a landscape. Chapter 3 shows how these realms are integrated and, if possible and relevant, how they are addressed.

The results presented in chapter 3 and especially section 3.2.1 – 3.2.3 refer to various levels, from national to field level and from a-biotic and biotic aspects to cultural aspects, via the socio-economic aspects of sustainable landscape management. In each section of 3.2.1 - 3.2.3 an introduction is given about how this realm is related to the other realms and thus how the disciplines (columns) are related to each other. Such an approach requires some flexibility from the reader. In order to give the reader some handles to pass through the results of this study, as presented in chapter 3, a short explanation and background of the different realms, 3.2.1 the a-biotic and biotic realm, 3.2.2 the social realm and 3.2.3 and the humanity realm, is given here.

The a-biotic and biotic realm (3.2.1): β sciences in column 1 and 2 of the Tableof the Checklist, viz.: environment and ecology

In line with the priority of physical survival, the first discipline or column of the presented Table of the Checklist (Table 3.1) contains the main targets and criteria of the landscape environment to warrant a sufficiently healthy and clean support to the landscape biosphere. The main targets and criteria of the webbed ecosystem of the landscape are given in the second column of the Table of the Checklist (Table 3.1). These targets and criteria should warrant food and fibre production of the landscape. The first two disciplines or columns of Table 3.1 (environment and ecology) draw mainly on the knowledge, methodology and approaches developed in β sciences' disciplines. The way they are organised here is derived from a focus on the physical survival of the landscape, including the population, which depends on the landscape for its survival. Thus, in the first column (environment), sustainable resource management is the leading theme and in the second column (ecology) the spatial and temporal connection or relationships between the diverse units such as species, bio-topes and landscapes, are the main issues. Diversity in coherence is the leading theme in the second column (ecology).

The social realm (3.2.2): γ sciences in column 3 and 4 of the Table of the Checklist (Table 3.1), viz.: economy and sociology

Where social survival of the landscape is the target of physical survival of the landscape, it is at the same time instrumental to the survival of the aesthetic and ethical dimensions of the landscape. So, the quality of the social environment of the landscape is the next priority of Maslov's triangle and is translated into the economic and social targets and criteria of the Table of Checklist (Table 3.1): economy (column 3) and sociology (column 4). The economic targets of column 3 of the Table of the Checklist refer to the flows of finances and services produced by and required for the sustainable management of the landscape. The use of money is seen as a social act that expresses the appreciation of commodities or services. The social targets of column 4 of the Table of the Checklist refer to the structures and procedures allowing those living in and in charge of the landscape management to participate in decision-making. Economy is regarded as the functional side of the social realm and sociology refers more to the status or recognition of the actors in charge of and participating in all various aspects of landscape management. The social realm is situated between the first and third realm. The social realm should balance the physical requirements in the context of limited resources (column 1 environment and column 2 ecology) with the immaterial qualities of cultures in the context of unlimited potentials (column 5 psychology and column 6 physiognomy and cultural geography).

The requirements of both major realms, β and α, are absolutely different for all individual people the original Maslow triangle refers to. They are also different for all European landscapes here at stake. Fair sharing of money and the possibilities to earn money is the leading theme of the third column (economy). Equal access to decision making is the leading theme of the fourth column (sociology). These two themes could facilitate the most favourable adjustments of the material and immaterial features of landscapes. Empathy for the landscape as expressed in monetary and social connections is crucial in this intermediate realm.

The humanity realm (3.2.3): α sciences in column 5 and 6 of the Table of the Checklist (Table 3.1), viz.: psychology and physiognomy and cultural geography

Both previous realms, the a-biotic and biotic and the social realm, are instrumental to the humanity realm, which refers to the most human aspects of human development. In this humanity realm, a difference is made between the subjective experiences of workers, inhabitants, visitors and governors of the landscape on one side and the objective features of the physiognomy of the landscape on the other. These disciplines are represented in columns five and six of the Table of the Checklist (Table 3.1) and elaborated in section 3.2.3. The fifth column refers to landscape psychology, which studies how people perceive and appreciate landscapes and landscape features. It acknowledges how different people can experience a landscape, as well as which features (common denominators), nevertheless, are commonly appreciated by most people. A differentiated landscape can be favourable to satisfy different sets of appreciation, while commonly favoured landscape features can provide baseline qualities. Thus, the subjective perceptions of people about the identity of landscape characteristics provide targets for the psychological discipline (column 5 of the Table of the Checklist). The main targets of column six (physiognomy and cultural geography) are derived from the characteristics of the landscape's sustainable features and its identity as defined by experts. Thus, column six is based on disciplines like landscape physiognomy and cultural geography, landscape planning and landscape architecture. The targets of these two columns together (psychology and physiognomy) should warrant the aesthetic and ethic dimensions of the landscape development, sharing with the development of people, which is and will be expressed and impressed in its past values and future potentials.

After the results of the study, presented in chapter 3, the performances of organic agriculture related to sustainable landscape management are discussed in chapter 4. This chapter includes a quick scan of case studies about agricultural farming systems and sustainable landscape management.

The next chapter, chapter 5, gives an overview of the possible uses and users of the checklist developed in this concerted action programme. Finally, the conclusions are given in chapter 6.

CHAPTER 2   RESEARCH METHODS

2.1   Outline of a unifying a concept to assess landscape management

In general, for the development of standards, it is important to clarify the objective(s) to be reached by using such standards. Here, the objective is to make a set of standards for the landscape management, which warrants the sustainable development of landscape. To implement the objective, it is necessary to clarify what is meant with landscape and sustainable development. Here, a definition of landscape is derived from Van Mansvelt and Mulder (1993) and Van Mansvelt (1997). Landscape is the land sculptured through human cultivation of the land, in its full complexity of agro-sylvi-pastural land-use, together with the establishment of the infrastructure like roads and waterways and buildings. It reflects the priorities for land-use as set by the population of the landscape, in the context of former national policy. These priorities, existing in various realms of society, together with the aesthetic appreciation and technological progress merge into the historical succession of landscape, in a co-evolutionary context of nature and humanity. This evolutionary succession reflects the development of the successive cultures of people, active and in charge of the land-use in each particular landscape. A certain polarisation between different forces within society has presumably always existed in the development process of culture. At one side, there are forces opting for what they perceive as progress, while at the other side, there are forces opting for what they perceive as care and respect for the established values of nature and culture. In a conflict between these two forces, they tend to point to each other's narrow minded and unacceptable selfishness (figure 2.1).

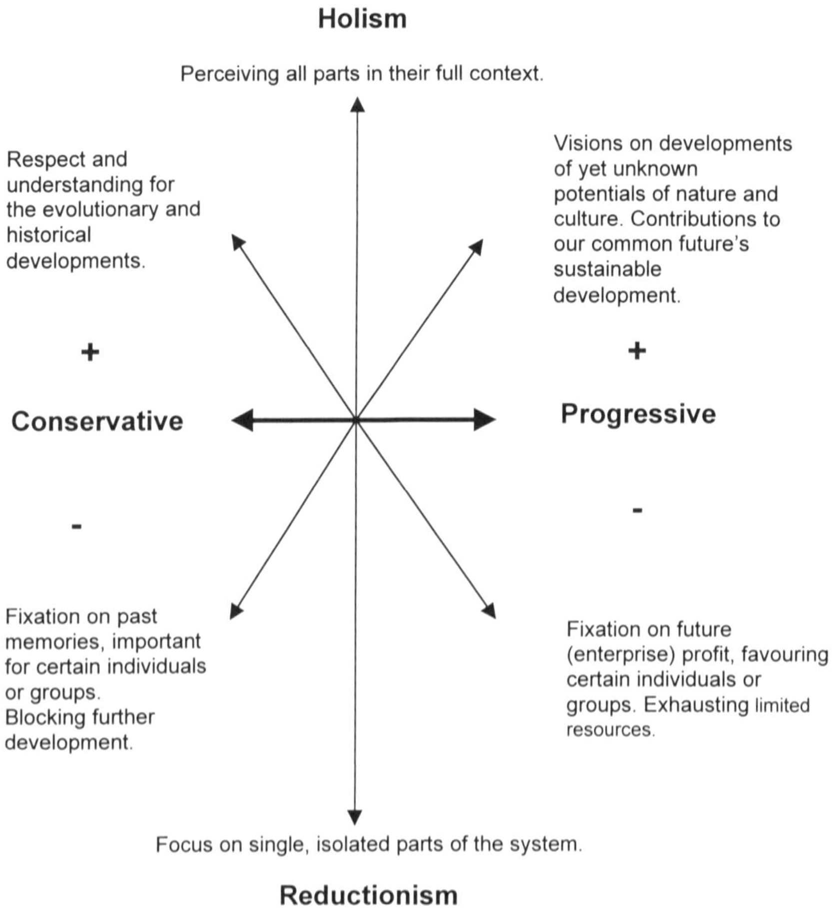

Holism

Perceiving all parts in their full context.

Respect and understanding for the evolutionary and historical developments.

Visions on developments of yet unknown potentials of nature and culture. Contributions to our common future's sustainable development.

+

+

Conservative

Progressive

-

-

Fixation on past memories, important for certain individuals or groups. Blocking further development.

Fixation on future (enterprise) profit, favouring certain individuals or groups. Exhausting limited resources.

Focus on single, isolated parts of the system.

Reductionism

Figure 2.1: Polarisation within society (horizontal) and the relation to a narrow or broad perspective.

The fixation of a more or less random site of species in a more or less random stage of evolution and history, having the sympathy of a small group in society, can be seen as the ego-centred bias of derailed conservatism. On the other hand, ruthless exploitation of irreplaceable values and resources of nature and culture, in favour of some larger or smaller industrial interest groups, can be seen as the ego-centred bias of derailed progress. Both kinds of derailing share a fixation on parts singled out of a wider context of spatial connections and temporal developments. Fixation on parts and moments disconnected from the whole, as may be incensed by radical reductionism, could enhance such derailing. However, both stands have their definite positive sides, which can be appreciated in their dialogue. A conservative respect for the world, in all its known and unknown beauty and wisdom created by evolution and history and perceived as requiring a most careful and retained handling, can be fully appreciated. This especially in view of the most disturbing, unforeseen and unwanted side effects of technologies, which are so widely applied. On the other hand, a progressive and courageous dedication to discover the entire world as yet unknown secrets, releasing all its potential benefits for mankind as yet untouched or unused to the detriment of people's happiness, can also be very well appreciated. This is especially in view of the continuing development of culture, consciousness and technology. Both positive sides of the conservative and progressive polar positions share the awareness of any object, part or moment in its spatial and temporal context.

Comprehensive or holistic approaches try to make sure that sufficient complementary steps in synthesis and contextualisation are made to complete all steps that have been taken in preceding research phases of analyses and reductionism. As pointed out by Van Mansvelt (1997), polar positions tend to root in attitudes, which are part of different paradigms, including more or less implicit perceptions about the relationship between humanity and nature. (Achterhuis, 1992; Bockemühl, 1992; Naess, 1975; Sheldrake, 1990). For instance, in certain perceptions humanity is seen as an evolutionary incident with humans as definite outsiders and exerting an inevitably negative impact on nature. Such a perception regards culture as incompatible with nature. In other perceptions humanity is seen as a co-evolutionary partner of nature, evolving from nature and inseparably dependent on nature, connected to nature and responsible for nature. Such different positions with various intermediates do also influence scientists in their perception and implementation of the ethical dimensions of their work and of science in general. It has become clear that the idea of neutral and objective scientific positions, which are presumed to be free of preconceived values, can not be supported anymore. Now, instead of the unfeasible pretension of neutrality, the strategy to mention explicitly the position(s) chosen, together with the choices related to that position, prevails. Being aware of this evolution, at least in applied sciences, it is concluded here that decisions related to society can and should not be delegated to scientists alone (Bosshard, in preparation). A political dialogue among all people concerned is needed. Therein, all discussion partners may consult those scientists they trust and need to underpin their arguments, while decision-making should proceed according to transparent and fair procedures. At least, that is the way it should work according to the rules set by the United Nations, presuming emancipated people in democratic societies (UNCED, 1992).

In this Concerted Action, the position is taken that humanity depends on living nature to survive and that nature nowadays depends on humanity for its survival as well. Moreover, underlying this mutual dependency for the common future of humanity and nature, the position is taken that such a common future builds on a co-evolution of nature and mankind. This co-evolution warrants a mutual compatibility of perspectives for a common survival in an ongoing development. Such perspectives could be challenged if mankind is perceived as definitely alien to nature. However, they fully comply with the perspective of the UN Agenda 21, building on the *Common future* of the Bruntland report (1987) and including the FAO recommendations about sustainable agriculture and rural development (SARD, 1991). All argue in favour of sustainable land-use and thus more or less to a sustainable management of the landscape.

European landscapes reflect human priorities, choices and interventions of the past, just like actual interventions will be reflected in the landscape of the future (Bockemühl, 1992). Landscape reflects the human motivations, as merged into social procedures of society's motivations. Human interventions are inevitably based on cultural perceptions, ideas and preferences of the actual actors in charge, made to interact in social structures and implemented as changes in the bio-physical conditions of the relevant localities. For example, concepts like 'natural' and 'non-nature-disturbing' management have changed considerably over the last decades, affecting policy and practice (Baldock and Beaufoy, 1993; Van der Windt, 1995). Similarly the concepts of system's efficiency is changing now, including more of what previously has been perceived as externalities (e.g. pollution, degradation). This affects, for example, which functions prevail in land-use, where they are to be allocated and to which extent, what are the criteria for good management practices, which targets should be reached and what impacts be accepted. In agricultural system thinking, such perceptual differences appear in their beneficial or detrimental impact on landscape, nature and environment (Baldock and Beaufoy, 1993; Pimentel *et al.*, 1995; Neugebauer *et al.*, 1996; Van Mansvelt and Mulder, 1993).

For a comprehensive overview of human motivation, the famous triangle of Maslow is chosen, which shows a clear mutual interdependence between human motivations, which can be perceived easily as contradictory or even as mutually exclusive. See figure 2.2.

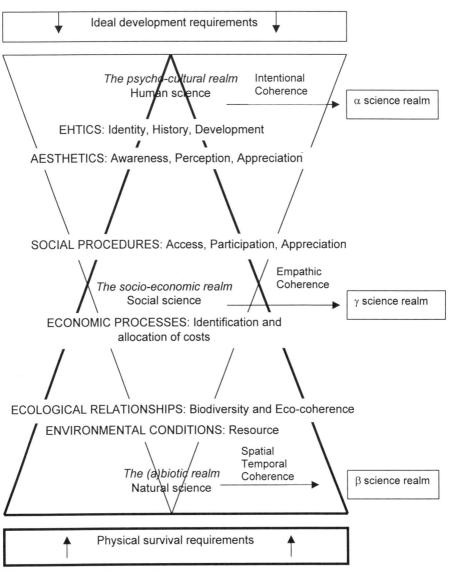

Figure 2.2: Adapted Maslow Triangle

Maslov's triangle is here subdivided into three stages with different human priorities at each stage. Arguing from bottom to top of the Maslov triangle, the basic priority of human motivations is getting sufficient water, food and shelter, in quantitative and qualitative terms, to warrant the physical and physiological survival. The next stage of the triangle is getting sufficient social functioning and status to warrant psychological survival. The top priority of human motivations then lays in the development of personal, unique and individual capacities that make every human being into a creation of its own ongoing biography. Now there clearly is an option to see a contradiction or paradox between the basic and the top motivation. This paradox is a polarity of a comprehensive system, with the basic priority as a prerequisite to release or effectuate the top priority. From that point of view, there is an upward spiral in human development, starting with an emphasis on food and shelter in the earliest childhood, adding emphasis on the elaboration of the next motivations during youth, adult life and seniority in an ongoing process of learning and self-development. However, the subsequent targets of the motivation for food, housing, salary and status can easily be used to compensate the next higher target and thus become a surrogate for the real motivation. This process is nicely phrased in sayings like "Eating to Live instead of Living to Eat" or "Working to Live instead of Living to Work". We are quite aware of the sensitivity of the debate on issues like "How much food is needed to survive?" "How big should houses be?" "How much status is needed to continue learning for self education?" "How egocentric is it to go for personal development?" However, these are the essential issues at stake in our society. They need positions to be taken and are actually affected by all decisions made. Our point of view is that the more explicitly decisions are made referring to the positions taken, the better it is for a democratic dialogue. Referring to the fact that the world is round, meaning that imperishable wastes can not be put outside the world and its resources are limited, meaning that they must be shared, it is relevant to be aware of the material impacts of the implementation of each of these positions. For instance, "What is the motivation for a certain group of people (country, enterprise, sector of society) to demand for scarce commodities and for non-renewable resources?" and "What is the sustainable development perspective of 20% of the world's population consuming 80% of the world's yearly available resources?"

Maslow's triangle translated to landscape

In an effort to translate the human motivations to objectives for sustainable landscape management, the landscape -in all it's aspects- is presumed to have a similar function for society as the human body has for an individual person: it carries and expresses society's development. Landscape needs appropriate food and shelter as the basis for it's physical survival, in order to produce food and shelter as the basis for society's physical survival. In such a landscape, the soil-ecosystem plays a crucial role as a most important interface between the a-biotic environment and living nature's hierarchy of ecosystems. The next stage of Maslov's triangle, the human motivation regarding a role and status in society, is translated in two ways: 1) the role and status that a landscape and its population have in the political context of a country, and 2) the need for the landscape biosphere to have sufficient social recognition from the local population to be maintained and managed in a sustainable way. The recognition of the role and status of landscape in society is partly expressed in monetary terms and partly in juridical ones like ownership laws, environmental laws, rules and obligations. Finally, the human motivation for self-development is translated to the motivation of a region to develop specific landscape and regional features, which create the character and identity of their own region in which they live and/or work.

Interesting is the fact that recently the post-second-world-war period of homogenisation, in which somehow everywhere and for every problem the most efficient and one-and-only solution has been applied, is followed now by a strong demand for diversification and quality building. Not only the common denominators are the leading issues any more. The demand for diversification and quality building is extended to and followed up by an additional trend to recognise, respect and develop the individual and cultural differences in a multicultural society. This holds for human individuals as well as for communities as for sites and regions. Thereby it should be stressed that multicultural is not only accepted as an inevitable fact of implemented human rights, allowing people to be as different as they want to be. In addition to the freedom of speech and religion rights side, multicultural can be seen as an enriching feature of our culture, allowing intercultural dialogue as an excellent tool for individual development and thus as one of the benefits of our times. Exactly the same reasons are valid for individualisation of regions and landscapes, which by appreciating their own geo-morphology, evolution and history, can always be seen in a wider context of a region having their own identity developed to the next stage of the triangle. The idea of this study is that the economic, social and ecological contribution of a region to society depends on the uniqueness of its quality, presumed that quality can be sustained and developed over time.

From the above presented considerations, it may become quite clear that agro-sylvi-pastural landscape is perceived not only as an expression of society's perceptions and preferences, but even more as an inseparable part of society's culture. It can be seen as a cover, sheath, coat or dress of society, just as a house can be seen as the cover a family lives in or as a coat or dress a person wears as an extended skin. They all, at a decreasing level of internalisation, are materials transformed by humans, individually and socially. Thus the landscape can be seen as a material part of society's body, expressing its immaterial identity in that landscape. This perception indicates the comparison of Maslow's human motivations to prerequisites for a sustainable landscape management. Nature's survival (sustainable biosphere development) is a prerequisite for society's survival (its sustainable development), which again is a prerequisite for human identity's survival (its sustainable development). The latter can be seen as an ultimate objective with the previous ones as prerequisites. However, in view of the earth's limited resources and the extreme inequity of their actual consumption (north-south), adaptation of the resources used for self-realisation to their long time availability is highly due (Daly et al., 1990; Lungren and Friemel, 1994).

To find common ground in the overwhelming range of values and criteria for sustainable land-use and landscape management, the sequence of human needs from Maslov's triangle is reflected in the approach used here. Starting from sheer survival (Maslov's first stage: food and shelter), via social integration (Maslov's second stage: belonging or acceptance and recognition or appreciation) to cultural development (Maslov's third stage: self-actualisation or identity development). These points are largely in line with the requirements phrased by the UN's Food and Agriculture Organisation (FAO) for its policy toward sustainable agriculture and rural development (SARD, 1991) and the UNCED's agricultural chapter of Agenda 21. There, food security (Maslov's first stage), employment, income generation and people's participation (Maslov's second stage) and human development (Maslov's third stage) are clearly addressed. Survival of the earth's biosphere is seen as the basis for sustainable food production and thus for societies' survival. In a Dutch policy paper on landscape and countryside planning, the full range of interests mentioned above is addressed with three E's: Ecology (the biosphere), Economy (the socio-economic sphere) and the Aesthetics (the sphere of humanities) (LNV, 1992). In the EU's agricultural policy, the same range of issues is addressed in some way or another (Baillieux and Scharpe, 1994; Baldock and Mitchell, 1995; De Putter, 1995; Giorgis, 1995).

An interdisciplinary approach

To achieve a science based and consistently sustainable landscape management, an interdisciplinary approach is a necessity, although it may obviously conflict with disciplinary approaches involved with sustainable development themselves. Disciplinary in-depth studies are concerned about more details, smaller scales and finer grained networks then interdisciplinary studies are. However, the reality of people's daily life in the rural areas, has to deal with how all the different aspects and functions of landscape, as studied by the disciplinary experts, combine and fit together in their world as a whole.

Within and between the various study disciplines relevant for the landscape (geo-morphology and environmental sciences, economic and sociological sciences, psychology and anthropology) professional languages, research methods, theories and approaches can differ considerably. In particular the way scientists from different disciplines usually look, ask and think may be experienced as incompatible with each other. This involves some 'not invented by us' attitude as well. To overcome problems of misunderstanding, based on different habits and points of view, awareness can be created among the different disciplines about the fact that they are all developed by rational human beings, trying to grip certain observations and trying to make sense of what they observe. In the end, the results should be applicable in today's real world. They should be communicable to and understandable for insiders as well as outsiders willing to face the research object on the spot. By sharing the observations and the way they could be analysed, explained or understood, disciplinary experts can meet on a high level of integration, as long as the applied research methods are recognised as instruments and not as major determinants for the research. In the latter case, instruments to meet targets would be inverted into instruments that become targets themselves. Thereby, an aspect of the whole, which is envisaged as the target, would become the whole itself, although being only an aspect of the whole, with a limited importance in that whole. Thus swapping the target for a tool to reach the target, a kind of spiralling process can start. (Bosshard, in preparation).

Looking to Maslov's triangle in figure 2.2 and the concept explained above, such a downward spiral happens for example if food, as an instrument for physical survival (stage one of Maslov's triangle), is taken as a substitute for social satisfaction. Another example how the downward spiral can happen is if a person's status is taken as a substitute instead of an instrument for personal development. In a similar way, when disciplinary interests prevail over the problems to be tackled, research can become too much method-oriented instead of target-oriented. Then any available method or a fancy research method or tool can become a must for any research to be taken serious, disregarding the actual appropriateness of that tool in view of the actual problem. Structural disciplinary interests of this kind can be found in scientific networks, but also in political and institutional ones. A strong focus on a disciplinary approach tends as of its own character to a strong competition between the disciplines, each claiming to produce the final solution of the problem.

However, with clear targets and awareness on the instrumentality of the prevailing research methods, common observations and shared perceptions, on a high level of integration, synergy between disciplines may be well possible. Besides that, if the origin of all scientific disciplines in the human capacities of understanding is acknowledged and the inevitable arbitrariness of decision-making is considered, interdisciplinary co-operation is very much enhanced. In the end it may then be possible that the human sense of wisdom, reviewing the relevance of all facts and rational arguments considered, determines decision-making in science, practice and policy. Not the results of any single scientific method is conclusive, but the human mentality is, as precisely the choice about the decisive method(s) chosen, is inevitable a human affair. Thus, the key for interdisciplinary co-operation lies in the human mind and mentality. Integration of the results of multiple disciplines will be found in the human mind and mentality and not in any discipline itself. By a conscious weighing process of the team, accepting the inevitably subjective character of each member's judgement, transparency of the decision making is enhanced. It is the challenge for co-operative and interdisciplinary judgement building, which becomes much more interesting then fighting questions about the best and ultimately decisive disciplines. Referring to the multi-functionality of landscape, interdisciplinary co-operation, as indicated above, seems the most plausible, feasible and promising approach.

An interdisciplinary approach and the Maslov theory are the methods applied in this study and will be further explained throughout the other parts of the study and especially in chapter 3 about the criteria for the ecological realm, the social realm and the psycho-cultural realm.

2.2 ON THE COMPLEMENTARITY OF INTERDISCIPLINARY HOLISM AND DISCIPLINARY REDUCTIONISM

To apply the checklist for sustainable landscape management, as presented in Table 3.1, in the way it is meant, it is important to be very clear about the status of such terms as parameters, values, criteria and targets. This is particularly the case in the context of the holistic approach that complements the reductionist approach and characterises the methodology of this work. Like knowledge on all details is important for the reductionist approach and reductionism for the knowledge on all fine details, so is holism for the knowledge on any object as a whole. The whole or Holon contains all details from which analyses of a certain (chosen) aspect of that Holon are derived. A definition of the 'whole' that is studied, assessed or analysed is a crucial part of scientific endeavours as mentioned above. Here, where the topic or subject of study is the landscape in its full multifunctionality, it may be obvious that no single discipline can cover all aspects, which have to be analysed. In order to study the agro-rural landscape management as a whole, a team of several disciplinary experts was needed and established (see chapter 2 Research methods). Moreover, to make their disciplinary work fruitful for the purpose as set, a shared perception of the landscape as a whole was crucial. By over and again asking one another within the team for the ideas and values encompassed in or behind the value of a parameter, the rationale of the use of that parameter and the target behind the criteria proposed could be identified and shared. Thus, a common perception of the landscape as a whole emerged from this iterative moving up and down from conceptual wholes to empirical data, passing all mentioned steps in between.

We notice that parameters are necessary to assess the landscape. However, at the moment that we come with any data on any parameter: tons of wheat/ha, mg NO3/l, red-list-species/surface unit, we still have nothing feasible in our hand. If we know how what we would like to reach (target value), how much we can accept (acceptable value) or how much we think could be within reach (feasible value) then we have something in our hand. Only when target values for sustainable landscape management are set for each of the parameters then targets can be used in a meaningful way. Depending on the character of the parameter, targets can be more or less quantifiable. The next question is 'why to set the target of a parameter at a certain value or level?'. Only if the level or value of a target is clarified e.g. tons/ha or mg/l, 'hard' figures set can be very 'hard' and suggesting scientific reliability. Otherwise, these hard figures set have no real meaning. Hard figures always need 'soft' bedding to get any significance at all. However, this soft bedding, also indicated as 'context', is mostly included in the self-understanding of experts about the subject of their expertise and thus only rarely explicit. Within interdisciplinary co-operation, discussing, talking, explaining and explicating is very important to warrant mutual understanding.

Talking about the question "why a certain parameter should meet a certain level or should stay within a certain range of values:", indicates that a parameter represents a certain criterion. Usually, a criterion has a higher level of integration than a parameter. For example ten red-list species per hectare could represent the criterion of naturalness of a landscape site or region. Another example is: two large cattle units per hectare, which could represent the upper limit of the carrying capacity of a certain bio-tope. Parameters are tools to measure or assess whether a certain research subject meets the criteria set for the values, which should be warranted or reached. Thereby it goes without saying that the parameter values will be different if they are applied at a Mediterranean lowland coastal area or a Scottish highland area.

However, criteria themselves have little meaning. Like naturalness of a landscape, cleanliness of a stream or water body, acceptability of income, they all figure within a wider context for which they stand as representatives. For instance, the quality of water can be different for swimming, washing and drinking, although the value of the parameter can be the same. Water that is clean enough for swimming is not necessary clean enough for washing, drinking or medical use of high-tech applications. Objectives that a certain study subject should serves, give the final meaning or ultimate value of a criteria. Moreover, the health requirements for drinking water for children or other vulnerable groups differ from adults with a strong constitution. This means that one target may have several criteria.

So there are also several targets a sustainable landscape management has to meet. For each target there are several criteria that can be used to assess the degree in which the management meets that target. Now for each criterion there are several parameters that can be chosen to assess the degree in which that criterion is met in a satisfying way. Then again, each parameter can have several values that satisfy the target set under various conditions. For instance, ten degrees Celsius can be warm during winter, cold during summer and moderate during springtime and autumn. Thus, along this line of arguments, there is a sequence of:

1. Targets that need to be specified in several assessable criteria ;
2. Criteria that need several specific assessable parameters;
3. Parameters that need several magnitudes set to serve validation of the assessed study subject;
4. Data to allow appreciation of the assessment of the criteria on their compliance to the set targets.

In this sequence the specificity and concreteness increase and end in usually much appreciated hard facts that are deemed to allow clear trade outs and other unambiguous discussions and actions. For instance, the cleanliness of the water can be indicated by its NO3 content, by the content of coli-bacteria and by the presence of certain fish or Gammarid species. The same holds for birds or orchids in grasslands, old and local varieties in plant or animal production and so on. This whole problem is well known, for example in environmental sciences and landscape ecology. However, it is a problem that must be over and again kept in consciousness to prevent disciplinary self-understanding to overrule the wider, more general interests. However, in this same sequence, the self-understood meaning of the "pars pro toto" being of an inevitably soft or paradigmatic character, increases with the increasing hardness of data. The desirability of ten or fifteen tons of wheat per ha or ten to twelve thousands litres of milk per cow per year can only be appreciated in a very particular context of related considerations. For example, the need for money for "the" farmers, the need for food for "the" people, the need for efficient feed-conversion, the maximal production of "the" cow's physiology and so on. The same holds for the production of any other commodity in the landscape. The more an object is singled out to become technologically transparent and open for modelling, the more difficult it becomes to keep an eye on its functional links with the context from which it became singled out. The functional context meant can be mainly natural but is increasingly acknowledged as a socio-economic or even as a cultural context.

It may be obvious that the more these contexts are put aside, overseen or neglected, the faster the technological progress on commodities that are singled out of their wider contexts. However, the time needed for environmental problems to arise may decrease in some proportion, together with an proportional increase of the time needed to restore the object or commodity links with their various environments, such as natural, social and cultural qualities or context. From these considerations, the following figure can be abstracted.

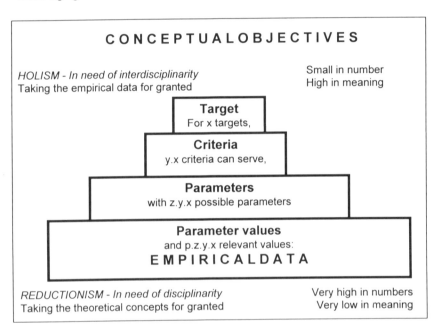

Figure 2.3: On the relationship between targets, criteria, parameters and the preferred values.

With this scheme in mind, the checklist can be more easily understood in its multi-scaled and inter-disciplinary character. When in the text explaining the checklist items, concepts and examples, theory and practice, larger and smaller scales in time and space will be addressed, the scheme as shown her might clarify what is going on and how the paragraph can be understood.

2.3 DISSEMINATION OF THE RESULTS

The successive achievements of the EU Concerted Action entitled *The Landscape and Nature production Capacity of Organic/Sustainable Types of Agriculture* (AIR3-CT93-1210) have already been disseminated through different outlets viz.:

1. Proceedings of the plenary meetings.
    Four plenary meetings have been organised during the Concerted Action. Proceedings have been produced out of the four meetings and were circulated among the participants of the concerted action, the EU-officer of the Concerted Action Programme and various other people who have shown interest in this subject. For details of the plenary meetings, see the proceedings written by Van Mansvelt *et al.* 1995a; Van Mansvelt *et al.* 1996a; Van Mansvelt and Stobbelaar, 1997; Stobbelaar and Van Mansvelt, 1994.

2. Journals.
    The majority of the papers produced during the Concerted Action and especially during the plenary meeting have been published in two special issues of *Agriculture, Ecosystems and Environment* (1997, 1998). Next to theoretical articles and articles about participants' own research results, these special issues also include the results of the quick scans performed on some farms in the landscapes of the participating countries during the concerted action meetings. The special issues of this journal show that they widen their scope from food and fibre production to landscape production aspects of agriculture.

3. A publication entitled *Checklist for sustainable landscape management*, published by Elsevier, 1999.
    The final results of the checklist and the whole framework of the Concerted Action are published as a document that may help farmers, policy makers, government and politicians to manage the development of agro- and forestry-landscape toward sustainability and socio-cultural appreciation. It can be used as a guide for agro-landscape production and as a framework for the follow-up and updating of the standards, criteria and parameters for sustainable landscape development. It can also be used as the basis for a communication tool for all stakeholders in a certain region within a process of regional identity development.

4. Dialogues and discussions with farmers and regional experts.
   Joint observations of the agro-landscape at farms and in co-operation with farmers and local and regional experts served as an important tool to raise awareness on the issue of sustainable agro-landscape development and to the dissemination of the concerted action approach in its process to produce the final checklist.

5. Dialogues and discussions with colleagues at conferences.
   Preliminary reports of the work so far done on the concerted action have been presented on the 11th International Scientific IFOAM Conference, titled *Sustainable agriculture for food, energy, and industry*, and organised in Braunschweig, Germany, June 1997. In October 1997, the final plenary meeting of the concerted action was invited to figure as a special workshop on the 25th anniversary of the WLO Conference. This conference has been organised in Bergen, The Netherlands and was titled *Landscape ecology: things to do*.

In November 1998, the framework and checklist will be presented at the 12th International Scientific IFOAM Conference, titled *Credibility of organic agriculture in the 21th century*, and organised in Mar del Plata, Argentine.

CHAPTER 3   RESULTS

3.1   INTRODUCTION

The results of the Concerted Action, AIR3-CT93-1210: *The landscape and nature production capacity of organic/sustainable types of agriculture*, are presented in Table 3.1. This Table of the Checklist is divided into three quality categories, viz.: the quality of the a-biotic and biotic environment, the quality of the social environment and the quality of the cultural environment. Within each quality category, two specific disciplines are distinguished.   The disciplines environment and ecology are distinguished to specify the quality of the a-biotic and biotic environment, economics and sociology to specify the quality of the social environment en psychology and physiognomy/cultural geography to specify the quality of the cultural environment. All these different types of disciplines are presented through six columns in Table 3.1 (the Table of the Checklist). In order to facilitate a sustainable management of a landscape, a comprehensive set of parameters and criteria has been drafted for each discipline and thus for each column of the Table of the Checklist. Thereby, an attempt has been made to define the parameters and criteria and to outline their importance (see section 3.2). Moreover, problems inherent to a non-sustainable management of the parameters and criteria are described. Indications are given on how to assess such problems. In order to increase the system's practical use, the techniques proposed for the assessment of the mentioned parameters and criteria are often of a qualitative and visual nature. Only when indispensable, more expensive and complicated, quantitative measures like (bio)-chemical or (bio)-physical analyses are indicated. If possible and relevant, for each of the parameters with respect to environment and ecology a desired range is suggested necessary to make sure they warrant that the goals of sustainable management of the rural landscapes are reached. These ranges of the values, however, are by no means universal. They vary regionally and locally in relation to the given natural conditions (e.g. soil type, climate, altitude, relief, etc.) and can also vary within the region's socio-economic conditions, as well as its human capacities and cultural values. The criteria and parameters of the socio-economic conditions are described in column 3 and column 4 of Table 3.1. Column 3 concerns the economic criteria and column 4 concerns the social criteria for sustainable landscape management. Finally, the criteria and parameters of human capacities and cultural values are described in the fifth and sixth column of the Table of the Checklist (Table 3.1), viz.: psychology and physiognomy/cultural geography.

After this introductory section (3.1), which contains the Table of the Checklist (Table 3.1), a detailed explanation and description of the criteria and parameters of the checklist are given in the next section (3.2). Section 3.2 is based on Table 3.1, which forms with its six columns the Table of the Checklist. The **main criteria** of each column are presented in bold and 14 inch letters and the **sub-criteria** are presented in bold and 12 inch letters. For each main criterion and sub-criterion a <u>TARGET</u> is given to serve the criteria and to give the reasons why they are mentioned as criteria.

Finally, to indicate ways in which the mentioned criteria can be measured, **_parameters_** are mentioned in bold and italics. Where relevant data could be identified, examples are given for magnitudes, limits or ranges in which we think these parameters should be kept in order to warrant the environmental qualities needed for its sustainable use. That is: to reach the set targets. The choice of parameters, their ranges and arguments are given and the choice of parameters is decisive.

In the elaboration of detailed targets, parameters and values of sustainable land use, the theories and practices of decades of organic and other environmentally friendly types of agriculture are taken into consideration as a considerable body of knowledge. As the main targets of those types of agriculture are widely in line with those of sustainable land use, it seems relevant to point to the "good farming practices" as developed in those circles of farmers and researchers. However, no explicit distinction is made in this study to any specific farming strategy. Examples of such strategies could be: High External Input Agriculture [HEIA], Low External Input Agriculture [LEIA], Low External Input Sustainable Agriculture [LEISA], Integrated Pest Management [IPM], Integrated Plant Nutrition Systems [IPNS], Environmental Friendly Farming [EFF], Organic Farming [OF], Biological Farming [BF], Ecological Agriculture [EA], Bio-dynamic Agriculture (BdA), Permaculture [PC], etc. Elaboration and inclusion of all these strategies would widely enlarge the framework of this study. Here, sustainable land use, including agriculture, is seen as an overall system of soil-based, multifunctional and primary production systems.

Conclusions about the checklist of sustainable landscape management are given in section 3.3. This section also includes an explanation of the synergy between the different columns of the Table of the Checklist (Table 3.1).

Table 3.1: Criteria for the development of sustainable rural landscape management

Quality of the (a)biotic environment		Quality of the social environment		Quality of the cultural environment	
1. Environment	**2. Ecology**	**3. Economy**	**4. Sociology**	**5. Psychology**	**6. Physiognomy / cultural geography**
Resource conditions	Biological relationships	Flows of finances and services	Participation procedures	Subjective regional landscape appreciation	Objective regional landscape identity
1.1 Clean environment • Fertile and resilient soil • Water quality • Air quality • Wild fire control 1.2 Food and fibre sufficiency and quality • Nationally sufficient and regionally sustainable levels of food and fibre production • Good food and fibre quality to match sufficient quantities 1.3 Regional carrying capacity 1.4 Economic and efficient use of resources 1.5 Sustainable, site-adapted and regionally specific production systems	2.1 Bio-diversity • Flora and fauna species' diversity • Bio-tope diversity • Ecosystems' diversity 2.2 Ecological coherence • Vertical coherence: onsite • Horizontal coherence: in the landscape • Cyclical coherence: in time 2.3 Eco-regulation 2.4 Animal welfare	3.1 Good farming should pay-off 3.2 Greening the economy 3.3 Regional autonomy	4.1 Well-being in the area 4.2 Permanent education of farmers 4.3 Access to participation • Farmer's involvement in activities outside the farm • Outsiders' involvement in farm activities 4.4 Accessibility of the landscape	5.1 Compliance to the natural environment 5.2 Good use of the landscape's potential utility 5.3 Presence of naturalness 5.4 A rich and fair offer of sensory qualities, such as colours, smells and sounds 5.5 Experiences of unity, like for example: completeness, wholeness and spaciousness 5.6 Experienced historicity 5.7 Presence of cyclical developments, for example growth cycles and the seasons 5.8 Careful management of the landscape, for example at the level of maintenance	6.1 Diversity of landscape components 6.2 Coherence among landscape elements 6.3 Continuity of land-use and spatial arrangement

3.2 EXPLANATION OF THE CHECKLIST

The following three subsections give a detailed explanation and description of the checklist for sustainable landscape management. Section 3.2.1 presents the criteria for the (a)biotic realm, viz: the environment and ecology (column 1 and column 2 Table 3.1). In section 3.2.2 the criteria for the social realm, viz.: economy and sociology (column 3 and column 4 of Table 3.1) are described and explained. The last subsection, 3.2.3, includes the criteria for the human realm, viz.: psychology and physiognomy and cultural heritage (column 5 and column 6 of Table 3.1).

3.2.1 CRITERIA FOR THE A-BIOTIC AND BIOTIC REALM: ENVIRONMENT AND ECOLOGY

J.D. van Mansvelt and D. Znaor

Introduction

In the development of sustainable rural landscapes, aspects and processes from the a-biotic or environmental sphere and from the ecological or bio-sphere, do play a most fundamental role. So they are of crucial importance in the selection of criteria needed for the assessment of sustainability. Processes taking part in the a-biotic compartments like soil, water and air, do indispensably influence relations in the bio-sphere realm of a landscape (living organisms and ecosystems) and vice versa. Actually, the whole biosphere depends on the conditions that the a-biotic sphere provides for. The other way round, living nature creates non-living nature as a product of its vital processes. Both spheres are subject to a wide range of interconnected complexes of physical, (bio)chemical and ecological processes, assimilation and dissimulation, anabolism and catabolism, growth, maturation and decline. They all are quite sensible to endogenous control and also to the systems' exogenous pollution, degradation, depletion and destruction, because the nature of these processes is very subtle and dynamic and moreover locally determined. These days, they are increasingly influenced by a number of deteriorating side effects of industrial and infra-structural activities on a global-level.

The sustainable management of landscapes, that obviously includes the prevention of the negative effects of human activities on specific elements of the a-biotic- and the bio-sphere, bio-topes or whole landscapes, requires careful and comprehensive planning and an attitude of good stewardship. Overall objectives of such a sustainable landscape management should include a clean, healthy environment and a bio-(genetic) diversity, which respects the valuable heritage of natural and cultural evolution. Here, with the word environment we refer to the a-biotic sphere: soil, water, air and energy, targeting at favourable conditions of the earth's non-renewable resources. The biosphere is referred to as ecology, meaning all relevant biological and ecological relationships in space and in time.

41

It is important to realise that all elements of the environment and the biosphere have not only complicated interactions among themselves as such. The elements are also very dependent on the extent into and the way in which they are perceived by the people in charge of policy decision-making. In the social realm, as will be pointed out later on (see column 3 economy and column 4 sociology of Table 3.1), the existing differences between various actors and groups in getting access to decision-making are highly determinative for the kind of decisions taken. If experts or representatives are invited to participate in any decision-making process then the question is "who decides on whom to be invited in or out?" The various (potential) actors or actor-groups may have a combination of different economic and power interests. Especially in the context of policy shifts, e.g. toward a more sustainable land use, the vested interests connected to the previous policies have to find and accept ways to accommodate. A crucial question within economics is "which strategy is economic to whom?" The fairly usual and short term interest of relative smaller private groups has to be exchanged for the long term interest of larger and more communal groups. This requires another level of considerations with a rather ethical dimension, indicated as the cultural realm. Therein, the personal perceptions of the involved values are addressed as well as the explicit differences in paradigms and related value systems that make people prefer short term over and above long term or vice versa. For instance, the standard for clean water is not only determined in an indisputable scientific way. Examples of questions related to the standard for clean water are: "how clean should the water be and *of* what substance (absence or "x" concentration of "y" substances)" and "how clean should the water be *for* what kind of uses: human consumption, bathing, washing, or trout farming?" The same decisions have to de made for standards on the necessary diversity of plant and animal species in a valley, on top of a hill or on a south or a north exposed slope.

Aspects filtering in from the social and cultural realm should not be disregarded or even denied in the decision-making process of the environment and biosphere values and qualities. The setting of sustainable management standards can only be done in a satisfying way if the existence, the background and the future perspectives of such value-system related perceptions are explicitly addressed. Here, satisfying means: long time functioning and thus positively recognisable in the landscape in retrospective. (See the psychological aspects in section 3.2.3 and the anthropological aspects also in section 3.2.3.) In our study, the presence of experts who are familiar with the practices and ideas of organic agriculture was found to be very fruitful. They added experiences on possibilities to merge food production and nature production in not yet widely acknowledged ways. This once again stressed the importance of considering facts in their proper contexts.

Compliant to what has already been mentioned about the complex interactions between the a-biotic and the bio-sphere, is that targets, criteria and parameters mentioned in column 1 (environment) of Table 3.1 and resource conditions are thought to be respected and included in column 2 (ecology and biosphere) of Table 3.1. Subsequently, they will be included as basic requirements in the considerations made in column 3 (economy), column 4 (sociology), column 5 (psychology) and column 6 (psychology and physiognomy/cultural geography). All these targets, criteria and parameters are compliant to the idea that survival of the biosphere is a prerequisite for a sustainable human development.

The quality of the a-biotic biotic environment for sustainable landscape management is represented by the criteria for the a-biotic and biotic realm, viz.: environment and ecology. These two groups of criteria are presented in the first and second column of Table 3.1.

Quality of the (a)biotic environment: Criteria for the (a)biotic realm

Environment		Ecology	
Resource conditions		Biological relationships	
1.1	Clean environment	2.1	Bio-diversity
1.1.1	Fertile and resilient soil	2.1.1	Flora and fauna species' diversity
1.1.2	Water quality	2.1.2	Bio-tope diversity
1.1.3	Air quality	2.1.3	Ecosystems' diversity
1.1.4	Wild fire control		
		2.2	Ecological coherence
1.2	Food and fibre sufficiency and quality	2.2.1	Vertical coherence: onsite
		2.2.2	Horizontal coherence: in the landscape
1.2.1	Nationally sufficient and regionally sustainable levels of food and fibre production	2.2.3	Cyclical coherence: in time
1.2.2	Good food and fibre quality to match sufficient quantities	2.3	Eco-regulation
		2.4	Animal welfare
1.3	Regional carrying capacity		
1.4	Economic and efficient use of resources		
1.5	Sustainable, site-adapted and regionally specific production systems		

1 ENVIRONMENT (Column 1)

Five main criteria have been distinguished with respect to the resource conditions, viz.:

1.1 Clean environment
1.2 Food and fibre sufficiency and quality
1.3 Regional carrying capacity
1.4 Economic and efficient use of resources
1.5 Sustainable, site-adapted and regionally specific production systems

1.1 Main criterion: Clean environment (a-biotic)

MAIN TARGET: A clean environment and long term availability of resources, allowing for the human and rural well being and development in a sustainable developing biosphere.

This main criterion is subdivided into the following sub-criteria:

1.1.1 Fertile and resilient soil
1.1.2 Water quality
1.1.3 Air quality
1.1.4 Wild fire control

1.1.1 Sub-criterion: Fertile and resilient soil

> *"The farmer is the guardian of the soil and of the country" (CEC-DGXI, 1993)*

TARGET: Prevention of soil degradation (pollution and loss of structure) and soil erosion; incentives for long term soil fertility improvement in rural or agro-sylvi-pastural production regions.
Note: Soil structure and fertility are the key issues of sustainable land-use. However, a minimal amount of natural erosion is an intrinsic and appreciable aspect of the nutrient recycling within the biosphere. The natural erosion of the mineral underground, caused by natural weathering together with plant-root and edaphic activities, underlies the biosphere development per se.

Parameters and desired ranges for a fertile and resilient soil:

1　　　Manure quality (C/N ratio)
2　　　Stocking rate (SR) matching the soil and the carrying capacity of the system
3　　　Anti-erosive belts and contour tillage (*indicative parameter, see explanation in the text*)
4　　　Soil cover (winter or off-season) (*indicative parameter, see explanation in the text*)
5　　　Crop rotation and crop mixture (*indicative parameter, see explanation in the text*)
6　　　Soil structure and organic matter content (*indicative parameter, see explanation in the text*)

1　　*Manure quality (C/N ratio)*

The C/N ratio indicates the ratio between the contents in organic matter of carbon and nitrogen. Manure with a high C/N ratio usually composes of non-easily decomposable compounds of old plant origin, such as cellulose, lignin, or waxes from straw, sawdust, chopped branches, present in plant-waste composts. Manure with a low C/N ratio contains a high portion of easily decomposable compounds of young plant origin and of animal excrement origin, such as manure, urine, sugars, hemi-cellulose, proteins, present in animal types of manure and urine. In general, manure with a high C/N ratio has a high and direct soil-feeding effect and a lower indirect plant-feeding effect, whereas manure with a low C/N ratio has a high and direct plant-feeding effect and no or little soil-feeding effect. In this context it should be stressed that the widely used term "organic fertiliser" is not very concise, as it covers a range from chicken manure and pure slurry (with very low C/N ratios), to old farmyard manure and compost (with high C/N ratios). It should be mentioned that manure slurries tend to have negative effects on the development of soil ecosystems through mineralisation and leaching. This is especially the case when their C/N ratio is low, their microbial transformation has not proceeded and their water content is high. For the sustainable treatment of organic manure see Godden and Pennickx (1997), MacNaeidhe (1997) and Saraptaka and Zilka (1997).

Desired range of C/N ratio of the soil

Depending on the actual soil situation and the addressed manure target, the C/N ratio will differ. As over-manuring of nitrogen has several contra-productive side effects, such as leaching, stimulation of plant diseases and discouraging of biological N-fixation, the C/N ratio of applied manure should in general not differ too much from that of the soil humus. This means that the desired C/N ratio is ca. 11.

The following Table presents examples of the C/N ratio's of various types of manure:

Table 3.2: C/N ratio's per manure type

	C/N ratio		C/N ratio		C/N ratio
Cow urine	2-3	Cow manure	25-30	Straw from	60
Pig slurry	5-7	(FYM)		oats, rye	
Cow slurry	8-13	Garden wastes	20-60	Straw from	100
Wet chicken	10	Tree leaves	30-60	wheat, barely	
manure		Mushroom	40	Sawdust	100
Ripened manure	10-20	compost		Fagus	230
compost		Straw from	40-50	Sawdust Picea	100-130
Kitchen wastes	12-20	beans		Bark	200-500
Mowed grass	12-15	and peas		Paper	

Source: Gottschall, 1984

2 Stocking rate (SR) matching the soil and the carrying capacity of the system

A clear link has been found between stocking rate (SR) or cattle density (CD) and soil-erosion (Mearns, 1996). More than half of the world's pasturelands are overgrazed and subject to erosive degradations (Leloup, 1994; Worldwatch Institute, 1988). However, also undergrazing causes problems (Leloup, 1994; Savory, 1988) Nowadays, any farmland or grassland area has a limited carrying capacity for two important aspects of animal husbandry: (1) for grazing or feed production and (2) for manure absorption. So, grazing has to be balanced between under- and overgrazing. Trampling, erosion and the composition of species have to be considered. The application of manure has to be balanced between soil depletion at one side and leaching and volatilisation from over-manuring at the other side. The primary productivity of the ecosystem and the percentage of net productivity which can be removed without affecting the restoring capacity of the ecosystem restoring capacities have to be considered if decisions about the sustainable SR have to be made. Mixed farms as well as grassland farms should not limit the SR to their amount of grasslands. Also the amount of arable land should be included, as these lands are necessary for the production of feed and straw and in need of proper manure. Proper in the sense of the feeding capacity of the ecosystem of the soil, that warrants a long term soil fertility. Green manure should be considered at the arable lands, which are important for human food production, in order to limit the need for animal manure. See also the parameter of crop rotation. For recent research on stockless farming see Cormack (1997); for beef and dairy see Ansaloni and De Roest (1997), Lampkin (1997), MacNaeidhe (1997), Mouchet and Boudier (1997) and Van Bol and Peeters (1997).

There are several visual indicators, which provide information about the stocking rate and the carrying capacity of the system. Examples of such visual indicators are the presence of palatable species and sods and pods, the disappearance of dung-pads, (edaphic activity) and the existence of erosion (ridges).

Desired range of SR

Based on long term experiences, the standards for organic agriculture indicate a maximum of 2.5 cattle units per hectare (2.5 CU/ha) in northern parts of Europe, 2.0 CU/ha in the central or continental part of Europe, and 1.5 CU/ha in the southern parts of Europe (IFOAM, 1996). However, the carrying capacity of the soil is higher in the case of mixed husbandry than for single species systems. Mixed farming systems allow explicitly for optimal management, because animal husbandry requires a minimal amount of feed and straw, which has to be grown somewhere; manure is required for the production of the required feed and straw. That mixed farms are not by definition a single farmers' farm is an important awareness, which will be elaborated in section 3.2.2 (criteria for the social realm: economy and sociology) (Baars, 1990; Gengenbach and Limbacher, 1991; Johnson *et al.*, 1951; Pimentel, 1993; Van Mansvelt and Mulder, 1993).

3   *Anti erosive belts and contour tillage*

Anti-erosive belts consist of permanent, graminaceous crops and/or shrub/tree-hedges, planted along the geographical contours to reduce and/or to prevent soil, water or wind erosion. Contour tillage means to cultivate the land parallel to the contour lines in order to prevent down-hill soil run-off through the tillage ridges. See also the ecological function in section 1.1.3.

There are several visual indicators, which provide information about erosive belts and contour tillage. Examples of such visual indicators are the assessment of the water or wind erosion effects, which can be measured by the amount of silt that covers the surface at the lower or down-wind parts of a field. Eroded parts or gullies of the field show lighter colours and exposed upper root parts.

Acceptable erosion

Maximally allowed annual soil erosion rate/ha depends on viz. should match the local soil formation rate. It is guessed that some 1 t/ha/year in temperate climate and 2 t/ha/year in the tropics are the maximal acceptable amounts of top-soil-loss.

Problem(s) of soil erosion

Soil erosion is one of the most burning environmental issues. During last 40 years, nearly one-third of the world's arable land has been lost by erosion and continues to be lost at the rate of more than 10 million hectares per year. Eroded soils have lower yields, lower fertility and reduced infiltration rates and water-holding capacity. Moderately, eroded soils adsorb 10-300 mm less water than non-eroded soils. Eroded soils contain less nutrients, organic matter and soil biota. They all have lower organic matter fraction and also less depth. Eroded soils cause shortage of basis plant nutrients compared to non-eroded soils (Pimentel *et al.*, 1995).

The average soil erosion rate varies from 30-40 t/ha/y in Asia to 17 t/ha/y in the USA and Europe (Barrow, 1991). However, these rates greatly exceed the average rate of soil formation, which is the rate of conversion of parent material into soils. Under tropical conditions the soil formation rate seems to be up to 2.0 t/ha (Pimentel, 1993), and about 1 t/ha/year in temperate climates (Troeh and Thompson, 1993). Between 200 and 1000 years are needed to form 2.5 cm of topsoil under cropland conditions. From these figures, 1t/ha/year can be provisionally set as a sustainable rate of soil loss (Pimentel *et al.*, 1995; Kabourakis, 1996).

Pimentel *et al.* (1995) calculated the economic costs of soil erosion at 8 $ t/ha/year. This means on a global scale $400 billion per year, which is $70 per person per year. According to Faeth (1993), the off-site social costs of erosion already amounts to $0.66-8.16 per ton of eroded soil.

4   Soil cover (winter or off-season)

In order to prevent nutrient leaching and soil erosion caused by water and wind and to provide habitat for beneficial organisms, theoretically the cultivated soils should be kept covered all year round (see also column 2 of the Table of the Checklist: ecology). However, in practice this may not always or everywhere be possible. For certain types of soil, such as heavy soils in severe winter climates where soil structure is improved through frost, or in arid and semi-arid climates where living plants evaporate much more water than bare soil does, exceptions from the above mentioned rule can be appropriate. In order to bridge over the fallow period between two main crops, cover crops for green manure and nutrient catching can grow. These are usually fast growing crops, which can be partially harvested for direct stable fodder. The cover or catch crop can be sown either as an associated-crop or as an under-sown crop to emerge after the main crop or as an after-crop that precedes the next main crop. Mulching the soil with plant residues and other organic material and leaving stubble after the harvest or late ploughing, also results in a certain type of soil cover. Although here, the focus is on agricultural land use, the same holds for silviculture or forestry.

The Soil Cover Index (SCI) is the percentage of the field or orchard covered by green manure crops, natural vegetation and cover-crop residues throughout the year. The SCI is equal to zero if the soil is completely fallow throughout the year and equal to one if the soil is completely covered by a cover-crop or cover crop residues throughout the year (Kabourakis, 1996; Vereijken, 1995). Monitoring can be done by visual estimations or the beaded string method (Sarantonio, 1991).

Desired range of SCI

The soil cover index ranges from 0 (completely fallow) to 1 (completely covered) (Vereijken, 1995). The optimum value varies, depending on soil type and climate. The optimum value of the seasonal and monthly indexes can also vary. Generally, winter months have rather high index (>0.60). The desired SCI in olive grooves is 0.5 (Kabourakis, 1996). It is important to link the SCI with the visible effects of soil erosion, which has to be prevented by the soil cover.

5   *Crop rotation and crop mixture*

Crop rotation is a valuable tool for many purposes, such as weed control, control of certain soil- and residue-born pests and diseases, maintenance of soil structure and organic matter and recycling of plant nutrients (Halley and Soffe, 1988; Van Mansvelt and Mulder, 1993; Vereijken, 1996b). Diversified crop rotation contributes to overall species and habitat diversity. It enables inclusion of restorative crops, such as soil fertility and soil structure building, and it ensures more equal distribution of labour and reduces the risk of economic failure. Biological N-fixation through bacteria and leguminous crops and the application of cover crops for erosion prevention, green manuring and nutrient catching, are also crucial aspects of crop rotation. Mixed cropping, pre- and after cropping are seen here as aspects of crop rotation.

When designing sustainable crop rotation, the following criteria should be taken into account (Van Mansvelt and Mulder, 1993; Vereijken, 1996b; Watson 1997; Znaor, 1996):

- The properties and features of each crop should be carefully considered for their multifunctionality and agro-eco-functionality in full crop rotation;
- A sustainable balance between annual and poly-annual crops (leys);
- A sustainable balance between leguminous crops (N fixation) and other crops;
- Soil coverage throughout the year (see parameter 4);
- To balance the water harvesting and water demanding features of crops in order to comply with the regional moisture conditions. A balance between water supply and water demand;
- The depths and intensity of rooting for soil structure improvement and feeding of the soil ecosystem. For instance, cereals, grasses and lucerne have a better rooting system than root-crops, tuber and bulb crops;
- The soil compaction should be linked with the required tillage per crop. Crops requiring only mowing have a better non-compaction value than crops that have

to be lifted from the soil during the summer and especially during autumn. See also the minimal tillage strategies;

- Crop's N take-off from soil reserves. Apart from leguminous crops, all other crops consume N from soil reserves. The nitrogen efficiency of crops and crop-species is to be considered in the perspective of soil mineralisation and humification throughout the rotation;
- The transfer of N to subsequent crops, which is based on (1) N residues in the soil after harvest, (2) N mineralisation from crop residues, (3) N losses from leaching and (4) denitrification. The higher the transfer, the better the crop. See also cover-cropping and nutrient-catching;
- The contribution of crops to the animal feed and straw production. This is important in mixed farming systems;
- The labour demand for e.g. preventive weed management, harvesting, tillage, etc.;
- The economic profitability per crop and per rotation or the net return per crop and rotation.

Vereijken (1996a, 1996b, 1998) emphasises the importance of the multifunctional crop rotation, which is the major method to achieve results desired in productivity, energy efficiency and soil fertility (incl. mineral balance). Besides, the multifunctional crop rotation supports plant species diversity and prevents/ reduces the use of pesticides. A good crop rotation (Vereijken 1996, 1998) should:

- Prevent the existence of a too high share of genetically and phytopathologically related groups (species and families) which share pests or plant diseases. The share per species should be <17-25% and per group < 33-50%. In other words there should be a sound balance in share of cereals, composites, umbelliferes, liliaceae, etc. include (autumn/winter) cover crops.
- Include some deep rooting crops. Iin this respect cereals, grasses and lucerne have better value than other crops (including green manuring crops), while the root, bulb and tuber crops have least significant here.
- Take care not to compact soil too much. Here the value declines from the crops moved in summer and autumn, lifted in summer and lifted in autumn.
- Have a good balance between high and low nitrogen uptake crops.
- Have a high N transfer to subsequent crops value, which is based on (1) N residues in the soil after harvest, (2) N mineralisation from crop residues, (3) N losses from leaching and (4) denitrification. The higher the transfer, the better the crop.
- Have a balanced N need, which is equal to N uptake minus N transfer (N need is net N input to be provided by manure or N fertiliser, and should be in the range of ≤ 2-3.

Desired range of crop rotation and crop mixture

In practice, a good crop rotation and crop mixture is usually a well-balanced compromise between soil improving crops and soil demanding cash crops. Examples of soil improving crops, which have a positive effect on soil fertility and structure building are leys, grasslands, legumes and green manure crops. Examples of soil demanding cash crops are cereals, tuber, bulb and root crops. The overall idea of sustainable soil management is that the farmer hands over a better soil to his successor than he got from his predecessor.

6   *Soil structure and organic matter content*

The term "*soil structure*" relates to the arrangement of primary particles of sand, silt and clay into ordered units (Halley and Soffe, 1988). However, it should always be seen in combination with the soil organic matter (SOM) which interacts with the mineral soil particles (Dekkers and Van der Werff, 1996; Hassink, 1995; Murata and Goh, 1997; Wander and Traina, 1996). A stable, erosion resistant soil structure is favoured by a high SOM level (high C/N ratio), a loamy or clayey soil texture and a high level of calcium. The SOM contributes to the cation exchange capacity of the soil, which facilitates the mineral nutrition uptake of the crops from the soil. It also contributes to better absorption of warmth by the soil during spring, which facilitates earlier crop development. Usually, soils with good structures have a better *porosity and aeration*, thus a higher moisture uptake after rains, and a better *water storage and retention* capacity during dry periods than soils with bad structures. At field level, good soil structure is important after sowing for the development of crops and rooting systems. Such soils tend to be easier to manage when they demand less horsepower and machine weight for sowing, tillage and harvesting. At the regional landscape level, water uptake and retention are of crucial importance for the water management. Especially in upstream areas, water uptake and retention prevent upstream erosion and landslides together with the prevention of downstream flooding. Finally, good management of the soil organic matter is important to feed the micro-fauna and -flora (edaphon), which in optimal conditions can importantly contribute to the suppression of pests and diseases or to remain within ecologically acceptable limits (Hoitink, 1989).

Soil structure can best be measured by the examination of a soil profile. Another method, which can be use to assess the soil structure is the so- called *spade diagnosis method* (Preuschen, 1985) Easily visible on-field signs, which attribute to a poor soil structure are (Halley and Stoffe, 1988):

- Poor or weedy patches, often accompanied by a hard and rutted or very cloddy surface layer;
- Overgrown, blocked an infield ditches;
- A non-friable plough layer with non-uniform colours. Rusty stained root channels and mottled structure faces, which usually indicate water lodging;

- The basis of the plough layer is not gradually merged, but has a sharp contrast to the subsoil. The latter is the result of a platy structured layer or plough pan;
- Under laboratory conditions. The soil organic matter is usually determined by wet oxidation (MAFF, 1996).

Desired range of soil structure and organic matter content (SOM)

Soils with a good structure have equally distributed roots are friable, and the structural aggregates do not show sharp edges (faces) when broken up. The soil is friable when the soil on the spade is not wider than the spade itself during soil digging. Soils with a bad structure have few small and/or very tortuous and non-continuous soil pores.

In Europe, more than 50 soil types have been distinguished. Each of them has its own characteristics, resulting from their particular pedo-genesis. Any type of soil is suitable for different types of land-use. Therefore, it is useless to prescribe a particular generic value for good soil organic mater. However, for arable land the soil organic matter content can be roughly indicated at >2.5%. For grasslands and horticulture a soil organic matter content of >4% is recommended. According to Vereijken (1995), the SOM should be within the range of 4% to 6%.

Soil organic matter problem(s)

Over several decades, the use of heavy machinery, careless soil cultivation, narrow crop rotations, and the omission or limited use of organic manuring in favour of mineral fertilisers will cause heavy losses of soil structure through soil erosion, even at the best soils such as Chernozems. The problem of lost soil structure has widely been countered by increased depth of tillage, which demanded heavier mechanisation, leading to increased compaction: a vicious circle. When, instead of crops, crop rotation and soil-feeding manure, mechanical soil conditioners are applied such as rotovators, the soil structure achieves only a temporary stage. This is an unsustainable situation of the soil. Such physical structures lack the stability of the organic soil structure, as caused by plant roots and edaphic activity. The latter are less easily destroyed by mechanical cultivation, water and wind movement, and livestock treading as compared to the former mechanical ones.

The pest preventing or suppressive effect of soil organic matter was found at the values of a SOM > 3% (Hoitink, 1989).

1.1.2 Sub-criterion: Water quality

TARTGET: Clean and healthy fresh water (groundwater and surface water), prevention of water pollution and water depletion, incentives for the long-time conservation of drinking water quality and water reserve volumes in the relevant rural or agro-sylvi-pastural regions.
Note:Most water pollution has its source in flow emissions from polluted soils. Wastewater effluents must be carefully considered. Overall, wastewater effluents

are less important. See for instance in the Danube Basin and The Netherlands (NEPP3, 1998; Vollenbroek, 1994).

Parameters for clean and healthy water:

1   Cattle units per hectare
2   Level and time of manuring (quantity per hectare per year)
3   Waste water treatment
4   Bookkeeping of minerals and additives
5   Bookkeeping of other potential pollutants
6   Water use and management

1   Cattle units per hectare

As indicated before in sub-criterion 1.1.1 (fertile and resilient soil), the stocking rate (SR) or stocking density (SD) is usually considered as one of the important factors of low-input-intensity farming systems. The SR or SD is meant to warrant minimal leaching of pesticide residues and nitrogen or nutrients from the soil into the surface or ground water. However, IEEP (1994) warns that data in stocking density statistics are not necessarily comparable, because different methods are used in different EU member countries for the calculation of the livestock units (LU's) on which the SR and SD are based. Different types of animals, such as cattle, sheep and pigs as well as different types of uses, like dairy and beef-stock, should be clearly differentiated for the feeding strategy. For instance, mainly roughage or concentrates. Moreover, a farm's SR as such does not refer to any seasonal or annual variations in stocking density per land unit, which is a crucial aspect of the actual effect of the animal husbandry on soil and water quality. As most EU Member States use other methods for the LU calculation than the one used in the EU legislation, this problem needs keen attention to warrant fair international competition among the farmers.

2   Level and time of manuring (quantity per hectare per year)

From the point of sustainable land-use, manuring is seen as a soil-fertility improving action instead of a means for waste-disposal. Therefore, time and place of manuring and quality and quantity of manure need careful attention. Manuring should not take place during the period that the soil-ecosystem can not sufficiently absorb, internalise and fix the applied manure or when the soil-ecosystem can not pass the derived nutrients from the applied manure to a growing crop at the time of the nutrient release. Particularly waterlogged or frozen soils should be excluded from manuring. To prevent the need for untimely manuring as a waste disposal, sufficient storage capacity should be ensured per farm or farming system for all types of animal manure. Usually this requires storage facilities for ca. 6 months to store manure in moderate climates. As far as restrictions on the place of application are involved, any type of manure, rock fertiliser, lime or other soil input should not be applied within less than 3 meters from any field boundary or 10 meters from any watercourse. This is particular the case for frozen grounds or waterlogged (FWAG, 1995). Standards for the manure quality should be derived from the soil's capacity to internalise or take up the particular manure in its ecosystem as 'feed' for the soil and crop. Besides the C/N ratio, also the aerobic microbe content and moisture have to be considered. Organic farming standards limit the amount of external nutrients and manure quantity to be imported at the farm, e.g. bought-in types of feed-stuff, manure, etc. This means that the quantity of manure at organic farms is limited by the organic farming standards. Considering the capacity of the soil to uptake manure, the allowed manure quantities are based on a permitted stocking rate of an average of 2 cattle units/ha. In the standard calculations of organic farming, 1 cattle units is equal to 0.7 dung unit. One dung unit contains 80 kg N and 70 kg P_2O_5. Therefore, the maximum amount of the nutrients, which can be purchased from outside the farm is 112 kg N/ha and 98 kg P_2O_5/ha (Fragstein, 1996; IFOAM, 1996; SKAL, 1997).

3   *Wastewater treatment*

On-farm wastewater mainly derives from household wastes, washing stables and milking rooms and manure storing- and silage-effluents often contain heavy organic polluters. Especially, animal husbandry needs special attention from wastewater management. Concrete floors or condensed clay layers for manure sites with ditches and containers to collect the liquid effluents are means to control water pollution. Appropriate application of the collected effluents on the land, eventually after treatment in halophyte-filters, also prevent nutrient losses in ground- and surface water. Apart from warranting minimal emissions from mineral nutrients, also the spreading of antibiotics, pesticide residues and microbial organisms from the farm into the water compartment should be carefully considered, because they are potential sources of human and other animal diseases.

4   *Bookkeeping of minerals and additives*

Ideally, sustainable land-use demands the balance between applied nutrients on farm soils and exported nutrients from farm soils via harvesting and sold products, should be close or equal to zero. This balance should be considered in the context of building soil organic matter as mentioned earlier under parameter 6 (soil structure and organic matter content) of criterion 1.1.1 (fertile and resilient soil). A positive balance sheet between input and output of for example N, P or K can be quite acceptable as long as leaching or evaporation happens within acceptable limits. This means that water- and air-pollution are minimal and internalisation of the nutrients in the stable soil organic matter is optimal. The indicated mineral balance sheet is not only important at farm level, but also at field level, in view of their contributions to the regional soil and water quality of the landscape.

The Nitrogen Available Reserves (NAR) is a parameter to estimate N leaching. The NAR is derived from the environmentally acceptable range of minimal N-content in soil reserves at 0-100 cm depth, at the beginning of the leaching period. According to Vereijken (1995), the NAR should be less than 45 kg N/ha for sandy soils and less than 70 kg N/ha for clay soils. The maximum concentration of NO_3 is 50 mg NO_3/l and of N is 11.3 mg N/l according to the EU regulations for groundwater. However, recommended or target values are 25 mg NO_3/l or 5.6 mg N/l (RIVM, 1991). Table 3.3 presents the quality target values for N and P in freshwater.

Table 3.3: Quality target values for N and P in freshwater

	Total nitrogen (mg N/l)			Total phosphates (mg P/l)		
	Ground water	Surface water		Ground water	Surface water	
		Large waters	Small waters		Large waters	Small waters
Limited value	< 11.3	<2.2	<1.5			
Target value	<5.6				<0.15 (a)	<0.08 (a)
Reference value						
• Clay/peat (b)				<3.0		
• Sand (b)				<0.4 (d)		
• Advise (c)				<0.01		

(a) Concentrations expressed in total-phosphate;
(b) Reference of total-phosphate concentration for soil quality;
(c) Advise from the Technical Commission of Soil Conservation for P saturated soils. Concentrations in ortho-phosphate;
(d) This norm reduced to 0.15 (Van der Werff et al., 1995).

The Dutch Ministry of Agriculture Fisheries and Nature Management (LNV, 1993) and the National Institute for Environmental Protection (RIVM, 1993) stated that the maximum acceptable surplus of nutrients in Dutch agriculture is 85 kg N/ha grassland. For arable land and the combination of grassland and arable land, the following figures are stated by LNV and RIVM: 65 kg N/ha arable land and 5 à 25 kg P/ha grassland and arable land. From 1995 onwards, the maximum allowed amount of applied phosphates in Dutch agriculture has been 150 kg P/ha grassland and 110 kg P/ha maize and arable land (Mol, 1993; RIVM, 1991). In order to comply with the quality standards for Dutch water, Bouma et al. (1997) assessed the critical threshold for N leaching to be 34 kg N/ha/yr.

Table 3.4 shows the target values for the maximum nitrogen and phosphate losses in agriculture set by the Dutch government.

Table 3.4: Target values for the maximum nitrogen and phosphate losses in Dutch agriculture

Sector	N-losses		P205 losses	
	Year 2000	Year 2008	Year 2000	Year 2008
Arable lands	150	100	35	20
Horticulture	150	100	35	20
Grass lands	275	180	35	20

Source: CLM, 1997

In the year 2000, the amount of lost phosphate should be reduced to 110 kg per hectare grassland and 70 kg per hectare other crops. This will result at national level into a surplus of manure of 30-40 million kg phosphate with an average surplus of 23 kg phosphate per ha. Comparing studies about nitrogen losses in organic farming and other farming systems show that organic farming already complies to the EU allowed emissions and that organic farming matches the EU target values (Boisdon and l'Homme 1997; CLM, 1997; Kloen and Vereijken 1997; Kovar and Krasny, 1995; MacNaeidhe, 1997; Van Bol and Peeters, 1997; Van Mansvelt and Mulder, 1993).

5   Bookkeeping of other potential pollutants

Apart from N and P bookkeeping it is important that also the use of other inputs, which might affect the environment, is recorded. For example, the use of pesticides such as, fungicides, herbicides, insecticides and nematocides and animal medicines such as, antibiotics, parasiticides, micro-nutrients and hormones need careful management. They may cause environmental pollution through spraying on the soil with subsequent emissions to water and air (Kovar and Krasny, 1995). As desirable parameters and values Vereijken (1996a) proposes to use:

- Pesticide index (pesticide inputs in kg active ingredients (a.i.) ha/year) which should be between 0 kg/ha/year (ecological farms) and 1 kg/ha/year (integrated farms).
- Environmental exposure to biocides (annual exposure of the fields and of the environment in biocide active ingredients), which should be equal to zero.

6   Water use and management

Next to the prevention of water pollution, the prevention of water depletion is also very important for sustainable landscape management, which includes land-use and agriculture. Together with the enhanced water-uptake and retention capacity of well managed soils, some additional strategies for good water management are required: (1) minimal use of irrigation, (2) prevention of water evaporation losses, (3) lowering the groundwater table and (4) soil salination. In general, when crops are encouraged to grow their roots more deeply, they will find the water they need and use it most efficiently. However, by superficial tillage, soil capillaries can be broken and prevents water evaporation from the soil. Crops with a high water demand should be limited in dry areas and crops with a low water demand should be limited in wet areas in favour of crops that ecologically fit into the region. The increasing demand for regional speciality products can be explained in the same way. Species can either fit to old, wet conditions or adapt to new, dry conditions. See for instance, wet and dry rice species. It should be clear that it holds for silviculture as well as for agriculture. Both should preferable be managed in an integrated, organic and ecological way at watershed scale (Mateu et al., 1997).

Furthermore, the canalisation of streams should be fairly limited in order to prevent groundwater depletion in the upstream and flooding in the downstream areas of the landscape. The meandering of streams causes considerable water retention capacity of a river basin and watershed and at the same time a huge landscape biotope diversity and climate control through trees along the riversides as wind-shelters, etc. See also parameter 3 (anti-erosion belts and contour tillage) under sub-criterion 1.1.1 (fertile and resilient soil). Vereijken (1996b) proposes to use the irrigation index (ratio of the amount of irrigation water used to the desired one) as a parameter for sustainable water use (should be <1).

1.1.3   Sub-criterion: Air quality

TARGET: Clean, fresh and healthy air in the countryside, prevention of bad smell emissions and volatile emissions of pesticides and residues, which affect human beings and the ecosystem.

Note: Much air pollution is generated from soil-emissions, manure or slurry and surface water emissions like volatilisation. But also direct emissions from pesticide sprayings and husbandry housing are sources of air pollution. However, a certain *intrinsic emission* from animals like breathing and flatulency and also their excretions of urine and manure should not be regarded as unnatural, but appreciated in the context of ecological nutrient recycling. By limiting the number of cattle units per -well sheltered- surface unit, a natural ecosystem buff can be warranted. See also parameter 2 (stocking rate matching the soil and carrying capacity of the system) and parameter 3 (anti-erosive belts and contour tillage) of criterion 1.1.1 (fertile and resilient soil) and parameter 3 (wind-shelter belts) of criterions 1.1.3 (clean and healthy air).

Parameters for air quality:

1   Ammonia emissions
2   Other emissions
3   Wind-shelter belts

For the time being, agriculture at large is an important source of direct bad smell, acidification, atmospheric pollution, emitting methane, nitrous oxide, carbon dioxide and other greenhouse gases as well as pesticides and their residues.

1   *Ammonia emissions*

Based on the present knowledge of the regenerative capacity of ecosystems in regard to soil and water acidification, the Dutch authorities for agriculture and environment agreed on a target value for soil and water acidification, which has to be reached in the year 2000. This target amounts to be 2,400 acidity equivalents per ha per year with a maximum of 1,600 acidity equivalents from N compounds. For the year 2015 the target is reduced to 1,400 equivalents with at most 1,000 equivalents from N compounds (LNV, 1993; RIVM, 1993).

2   *Other emissions*

Also for the clean air, bookkeeping of the potential pollutants is very important. In particular, the use of biocide sprays should be well recorded, with the aim to minimise their use (drift and residues affecting not only the fields of the applicants but also their neighbours' fields (environmental compartments). See also parameter 5 of sub-criterion 1.1.2 (Water quality).

3   *Wind-shelter belts*

Wind-shelter belts and meadows function as bio-geo-chemical barriers to control the spatial spread of many pollutants and inorganic ions. They also control the leaching of nutrients, which in mosaic landscapes is 2-3 times less than in non-mosaic ones (Ryszkowski, 1995). Moreover, they can store about 20-60 mm (i.e. 20-60 l/m2) more water than open and uniform cultivated fields (Molga, 1986). The saving of water in rich sheltered regions can be 40 mm (i.e. 40 l/m2) more during hot and windy periods than plain fields. The potential evapotranspiration of rich sheltered areas may decrease by 34% compared with plain fields (Ryszkowski and Kedziora, 1987).

Moreover, wind-shelter belts and meadows offer a niche for beneficial insects and birds and have a recreational or pleasure value. For example, many predator species, which live at the farm fields, migrate between favourable and unfavourable habitats within the agro-landscape. Therefore, wind-shelter belts like bug-banks, grassy field margins, woodland and hedges are of crucial importance for many species which need a cover for over-wintering (Sotherton, 1985; Wallin, 1985). It is argued that winter crops and winter-sown cereals may enhance predator survival next to catching nutrients, preventing soil erosion and serving as green manure and animal feed (Booij and Noorlander, 1992). See for more details column 2 (ecology) of Table 3.1 (Table of the Checklist).

1.1.4   Sub-criterion: Wild fire control

TARGET: The prevention of uncontrolled wildfires, which damage the ecosystems. Basically, wildfires are a natural tool for bio-tope diversity management. Wildfires help with the renewing of the vegetation by creating well-measured clearings that allow annual and perennial species to spread and woody species to re-grow (Etienne, 1996; Silva Pando and Gonzales Hernandez, 1992). However, especially in the Mediterranean areas, wildfires and semi-wild fires have been misused to create clearings for building houses, villages or industries in nature conservation areas. Also in arable areas created burning straw from cereals huge losses in organic materials, which could have been used for animal husbandry purposes, such as bedding and manure absorption in farmyard (Goldhammer and Jenkins, 1990).

1.2 Main criterion: Food and fibre sufficiency and quality

MAIN TARGET: Providing sufficient amounts of good quality food and fibre to warrant sustainable human development of the global society. The basis for good and sustainable food supply is considered for regional self-sufficiency in staple food. Regional self-sufficiency can well be completed with sufficient interregional exchange of additional food products, like fresh or upgraded products that can not be sustainably produced in the region itself. Obviously, the concept of regional self-sufficiency should neither be over-stressed nor under-emphasised.

Note: As has been stressed before, it is important to realise that the criteria and parameters mentioned here presume that the before mentioned parameters are sufficiently well met to warrant their capacity to support the criteria and parameters of the ecological criteria (column 2 of Table 3.1). This is of special interest, as populations can no longer migrate away from a desertified area to occupy and exploit a next area. As the earth is round and human rights are meant to warrant equal access to the earth's resources, the roll-of of problems to elsewhere and later is not an option anymore.

From this main target the following sub-criteria are derived:

1.1.1 Nationally sufficient and regionally sustainable levels of food and fibre production

1.1.2 Good food and fibre quality to match sufficient quantities

1.2.1 Sub-criterion: Nationally sufficient and regionally sustainable levels of food and fibre production

TARGET: The basic idea is to find a regional balance within a national and supranational context between the carrying capacity of the region's environment and the population, which have to be nourished, clothed and provided with fibres and fuels. Thus provided with regenerative resources. To prevent that populations and food processing get too much centralised and excessive, often going along with rural degradation, must be counteracted (Hemalata *et al.* 1997; Sethuraman and Ahmed, 1992).

Parameters for nationally sufficient and regionally sustainable levels of food and fibre production:

1 Minimal nutrient requirements per capita, derived from the World Health Organisation WHO)

2 Required area for sustainable agriculture

3 Level of integration of land for food production and land for nature production

1   *Minimal nutrient requirements per capita, derived from the World Health Organisation*

Modelling the demand for nutrients at a global level requires a considerable number of decisions about presumptions, which have to be chosen for incorporation into the model. On the one hand, there are several international studies at global level, which mention that feeding the global population is highly at risk. It is argued for example, that to feed every human being with sufficient food and a diverse diet requires 0.5 hectare arable land per capita, while only 0.27 hectare arable land per capita is available (Van Dieren, 1995; Lai, 1989; Pimentel *et al.*, 1995). On the other hand, there are several studies at national levels, concerning European countries, which state that *even* when the whole particular country would shift to organic agriculture, the whole population of that particular country can be well nourished (Van Mansvelt and Mulder, 1993).

A crucial issue in the global nutrition debate is the demand for animal proteins in the human diet (Van Mansvelt, 1997). Another issue is the minimal daily requirements of nutrient intake which, interestingly, has been increased over the last decades. This increase of required nutrient intake is especially interesting in view of the massive over-consumption in the rich countries over the last decades. A third issue, that needs special attention, is the considerable losses of nutrients in the food chain from field to table. Also food losses in transport, storage, transformation and other post harvest losses should not be excluded.

2   *Required area for sustainable agriculture*

In the context of national sufficient and regional sustainable levels of food and fibre production, sustainable land-use should include enough area for sustainable agriculture to allow for a region's proportional contribution to the national and eventually international food supply.

3   *Level of integration of land for food production and land for nature production*

Irreversible over-exploitation of any natural resources within any region limits the perspectives for physical survival of people's next generations as well as that of the biosphere. Therefore, a well-balanced and ecologically appropriate location and integration of lands for nature production and for food production are necessary. See also parameter 1 of sub-criterion 2.1.2 (Bio-tope diversity).

1.2.2   Sub-criterion: Good food and fibre quality to match sufficient quantities

TARGET: Good quality food is needed to warrant a long-term and sustainable physical survival of the consumers.

Note:In this food quality assessment, the absence of health-stressing additives is not the only important feature. Just like (organic) agriculture is not 'good' only if it refrains from the use of pesticides and mineral fertilisers, nutrition is not necessarily 'good' if it does not contain 'bad' substances like nitrates, pesticide-residues, hormones or antibiotics. The presence of positive qualities such as nutrients, tastes, colours, structures, vitality, storage and capacity, is most crucial.

Parameters for good food and fibre quality to match sufficient quantities:

1   Self-balance in physiology of human organism
2   Good sensorial and nutritional qualities
3   Regionally specific quality

1   Self-balance in physiology of human organism

As (organic) agriculture is only 'good' if it manages to be well-balanced and relies on the synergy of agro-ecosystems' self-organising capacities, food might be 'good' if it facilitates the human organism to keep its physiology self-balanced. That means, in a dynamic and healthy balance of a multitude of parallel and opposing anabolic and catabolic activities to warrant the well-being which characterises human health experience (Schad, 1993; Tape, 1992; Van Vliet, 1998).

2   Good sensorial and nutritional qualities

Together with the optimal nutrient content, the overall nutritional quality of food is also dependent on good sensorial and nutritional qualities.

3   Regional specific quality

Here regional specific quality comes in as an issue that is worth to be considered as a positive asset. World wide, monotonous and anonymous or general food quality with an unspecific taste and structure is not as widely appreciated as has been presumed. Speciality products and 'appelation regionale' gain more and more appreciation and importance (Kiley Worthington, 1993; Meier Ploeger and Vogtmann, 1989; Poldervaart, 1996).

1.3   Main criterion: Regional carrying capacity

MAIN TARGET: The sustainable management of a landscape has to be sufficiently compatible with the carrying capacity of the natural resources in the region.

Note: Where main criterion 1.2 discusses the sustainable levels of food production in the context of sustainable land-us, main criterion 1.3 (regional carrying capacity) indicates that the sustainable management of a landscape has to be sufficiently compatible with the carrying capacity of its natural resources: soil, water and air. If landscape management is not compatible wit its natural resource, then irreversible degradation of the landscape ecosystem will take place. Whereas the discussion in criterion 1.2 (food and fibre sufficiency and quality) focuses on the importance and parameters for a sustainable ecosystems management, criterion 1.3 focuses on the sound environment as the basis for the sustainable biosphere development. In criterion 1.1 (fertile and resilient soil) the discussion already started about the need to prevent soil erosion as the ultimate requirement for sustainable landscape management.

Parameters for regional carrying capacity:

Several studies about the regional carrying capacity are available, e.g. Harris (1996), Sage amd Redclift (1994) Van Pelt (1993) and UNDP (1991). The development of good and relevant parameters for the regional carrying capacity needs further elaboration and research.

1.4 Main criterion: Economic and efficient use of resources

MAIN TARGET: Optimal economic and efficient use of resources in order to warrant the availability of resources for future human and biosphere generations and based on sustainable land-use development. However, optimal use of the *limited non-renewable* fossil and natural resources is unsustainable in the long term. This main criterion should be considered in addition to the before mentioned well treatment of the renewable resources of nature (see criteria 1.1.1 – 1.1.3).

Note: In the previous criteria, the efficient and multifunctional use of the environmental compartments has been addressed in the context of 'the earth is round awareness'. Nutrient balances and nutrient recycling have been mentioned as crucial issues on several levels and for several aspects of agro-silvi-pastural land-use including fruit, vegetable, herb and flower production. What remains to be mentioned here, is the domain of a 'most sustainable use' of the non-renewable resources, which are only available in clearly limited amounts.

Parameters for the economic and efficient use of resources:

1   Resource efficient energy management
2   Minimally required input of non-renewable energy
3   Dependence on non-renewable energy sources
4   Net yield from external non-renewable inputs

1   *Resource efficient energy management*

In agriculture, energy efficiency can be defined as the ratio between the energy equivalent of yield and the energy equivalent of external as well as internal inputs. According to Vereijken (1996a), this ratio should be higher than 10. However, various modern and high external input technologies in agriculture are much lower (Alfoldi *et al.*, 1995; Bujaki *et al.*, 1995; Greenpeace, 1992; Lünzer and Kieffer, 1992).

From various publications, like Defrancesco and Merlo (1996), Hueting *et al.* (1992), Lundgren and Friemel (1994), Tellarini *et al.* (1996) and many others, it becomes clear that the cost of fossil energy has been kept so low as if resources are unlimited. However, labour costs have been increased, giving strong incentives to swap from human to mechanised labour, as if the human labour capacity has been the limiting factor within a world with a fast growing population. For the rural landscape development, this means land-flight and urbanisation, which again resulted in a hard to reverse degradation of the rural landscape. To redirect land-use towards a sustainable rural development, which would result in the re-establishment and sustainable management of appreciated landscapes, also needs the redirection of energy- and labour-price policy. See also column 3 (economy) and column 4 (sociology) of Table 3.1 (Table of the Checklist). In general, when talking about resource depletion, it should be kept in mind that the combustion of fossil energy goes along with the acidification and polluting emissions into the air, such as various C, N, S and P compounds and thus add to the negative environmental effects. See for recent data Halberg (1997), Leake (1997) and Mouchet and Boudier (1997).

Vereijken (1994) indicates that the energy efficiency of sustainable farming systems should be >10 (the ratio of the energy equivalent of the yield to the energy inputs).

2   *Minimally required input of non-renewable energy*

By measuring the energy efficiency, a difference can be made between renewable and non-renewable energy. The latter is the issue of this criterion and refers to the environmental aspects of the landscape management. The actual solar energy is the basic natural source of energy in agro-silvi-pastural land-use. Therefore, the main clue for land use should be to optimise nature's energy fixing capacity. This should not be done in a restricted model, which maximises only single crop production, but – as indicated before- in an agro-ecosystem management context, because for example soil-fertility and soil-structure-building are intrinsic products of crop production. These products need their own proportion of solar energy, fixed by crops, in order to feed the soil ecosystem (edaphon) with energy for the transformation of crop and manure residues into relevant kinds of humus. Also biological N-fixation and adequate mineralisation require soil microbial activities, which demand their own portion of solar energy (Nijland and Schouls, 1997). See also sub criterion 1.1.1 (fertile and resilient soil) of Table 3.1 and the parameters for soil erosion and soil structure.

This parameter focuses on the minimally required input of fossil energy at farm level for activities such as tillage, sowing, harvesting, drying, cleaning, packaging and storing. Animal husbandry, heating, aeration for climate control, and manure management like watering and drying are also relevant.

3   Dependence on non-renewable energy sources

Next to the strictly on-farm energy efficiency, it is important to screen the whole agriculture-chain, ranging stepwise from the farmer's field to the consumer's table. Here, in particular the energy needed for transport and transformation is getting into the picture. Both are relevant for farm-outputs as well as farm-inputs. Famous examples are the fossil energy content of the industrial mineral N-fixation versus the biological N-fixation in the field or the fossil energy content of heated and lighted greenhouse products ("hors sol") versus the fossil energy content of products from the field. However, vegetables from the open field that are out of season and imported for other climatic regions may contain considerable amounts of fossil energy as well.

4   Net yield from external non-renewable inputs

Another parameter, which can be used to calculate the energy efficiency at any level of the agriculture-chain is to focus on the net yield or output of products in weights or energy contents and the per unit non-renewable energy input. The so calculated ratio can be applied on any level from the farmer's field to the consumer's table, presuming that relevant data are available.

1.5   Main criterion: Sustainable, site-adapted and regionally specific production systems

MAIN TARGET: In line with the previous criteria, the target of this criterion is to encourage local or regional specific production systems that warrant and support the necessary environmental conditions for the development of the ecosystem and bio-diversity. The development of local and regional specific production systems should be in favour and going together with the development of the ecosystem and bio-diversity.

Note: The first three sub-criteria, 1.1.1 (fertile and resilient soil), 1.1.2 (water quality) and 1.1.3 (air quality) focus mainly on the farm and field level. The main criteria 1.2 (food and fibre sufficiency and quality), 1.3 (regional carrying capacity) and 1.4 (economic and efficient use of resources) emphasis on and merge in the landscape of the region. Here at main criterion 1.5, we look at the management of the agro-sylvi-pastural production systems, which determine a lot of quality and development perspectives of the rural landscapes. Each parameter for this target and criterion is an extension of the previous one. These parameters also anticipate to the next columns of the Table of the Checklist (Table 3.1), from column 2 (ecology) to column 6 (physiognomy/cultural geography).

Parameters for sustainable, site-adapted and regionally specific production systems:

1	Locally adapted farm management
2	Cultivation of local crop and animal species
3	Production of regionally speciality products

1 Locally adapted farm management

It should be stressed again here, that all general farming methods and techniques do require considerable local fine-tuning to make them serving the purpose they are meant to serve. Soil climatic conditions, the steepness and exposure of slopes, the agricultural history of the farm, and the planting and aquatic infrastructure, they all influence decision-making at farm level for the best time to work and the best techniques to use. Crop rotation, stocking rate, tillage, grazing and mowing, manuring, and irrigation, they all need site-specific fine-tuning to warrant both optimal environmental conditions and optimal harvest, preferably assessed at agriculture-chain level instead of most-profitable-crop level. Anticipating already on column 2 (ecology) and column 4 (sociology) of Table 3.1, it should be emphasised that farmers have to be encouraged to evaluate and trust their own decisions for the improvement of their farm activities. The farm activities should be improved in such a way that they fit to their competence, preferences and the site conditions of the farm. This means that in addition to the general knowledge, farmers must be encouraged to elaborate their specific and individual knowledge. See also main criterion 4.2 (permanent education of farmers) of Table 3.1.

An important issue is that farming should not degrade the environment because of the application of technologies that might be useful somewhere else, but not on the actual farms conditions. Apart from general indications, simple straightforward environmental parameters, exceeding the parameters mentioned in other criteria, can not easily be found for the assessment of the degree of local adaptation. However, some relevant parameters are presented in column 2 (ecology) of Table 3.1.

2 Cultivation of local crop and animal species

Valuable European landscapes have their genesis in centuries of local farmer's populations developing animal and plant species which, by historical default, had to be and were indeed optimally adapted to local low external input and nutrient recycling conditions. Mostly, they were also rather resistant to the local conditions, including diseases. However, with the growth of traffic and technology options, being heavily facilitated by the governments, research and industry, there was a strong trend to 'modernise', 'rationalise' and move away from 'old fashioned habits'. Modern species, supposing to be of world wide profitability, have been selected and improved in order to be applied all over the world (Anonymus, 1991; Audiot and Flamant, 1992; Hobbelink and Thompson, 1993)

So far, little attention from the financial, political and scientific site has been paid to the upgrading of local breeds in line with the ongoing development underlying their own origin. This contributed importantly to the loss of bio-diversity, which is closely linked to the agricultural activities of our predecessors. If the identity of regional landscapes is emphasised in line with the previously mentioned targets for a sustainable landscape-management, then the reconsideration of breeding strategies is a complying policy (Haynes, 1994; Temirbekova and Van Mansvelt, 1998). Apart from the valid argument that pleads for the in-situ conservation of local varieties, the idea that the development of breeds and species toward farmers' goals could continue, does fully fits into the recent perceptions of nature conservation and genetic and bio-diversity protection. Also the objective to fix any natural habitat or region into any appreciated stage of history (e.g. back to the fifties or the beginning of the century) has given way to the objective of site specific *developments*. Such a strategy warrants a relatively natural and healthy development or at least warrants one of the possible developments that largely fit the ideas of natural succession processes. From this point of view, the parameter about the existence of local plant and animal species in farm and landscape management is a reliable parameter for this issue. The parameter refers clearly to all aspects of agro-silvi-pastural land-management at large. Moreover, if local plant and/or animal breeding activities exist within the region, then this parameter would clearly support the previous parameter. The support and involvement from the farmers' local and regional communities for such breeding activities, indicates that the social carrying capacity of such initiatives does exist. (See also column 4 (sociology) of Table 3.1).

3   *Production of regionally speciality products*

In addition to the use of regional crops, varieties and breeds, regional specific processing such as cheese, bread and bakery, drinks, meats, etc. may contribute to the onsite conservation of species and at the same time support the economic prosperity of the farmers and their region. See also the vertical coherence in main criterion 2.2 (ecological coherence) of Table 3.1 and further in column 3(economy) and column 4 (sociology) of Table 3.1. This aspect obviously indicates that special attention and promotion is required for regional products, which fit both, the ecosystem and the market.

2   ECOLOGY (Column 2)

Four main criteria have been distinguished with respect to the biological relationships, viz:

2.1   Bio-diversity
2.2   Ecological coherence
2.3   Eco-regulation
2.4   Animal welfare

2.1 Main criterion: Bio-diversity

MAIN TARGET: To safeguard a sustainable development of the regional landscape biosphere diversity, within the context of a well-structured and well-cultivated and regional and supra-regional network of ecosystems.
Note:The biosphere network of ecosystems has a number of functions towards the a-biotic environment, which it supports and depends on, like soil, water and air (see the clean environment criterion of Table 3.1). It has also a number of functions towards the human society, which it supports and depends on, like the socio-economic and cultural environment.

This main criterion is subdivided into the following sub-criteria:

2.1.1   Flora and fauna species' diversity
2.1.2   Bio-tope diversity
2.1.3   Ecosystems' diversity

2.1.1   Sub-criterion: Flora and fauna species' diversity

TARGET: To safeguard the continuing and sustainable in-situ development of a valuable species bio-diversity and genetic diversity. This refers to un-cultivated or natural and cultivated species, bio-types, bio-topes and valuable and appreciated landscapes. The idea is to warrant per region and per bio-tope, the existence and where relevant the continuing development of characteristic plant and animal species, which requires a minimal abundance of plant and animal species. Species can be rare, threatened or still common and can be characteristic for a natural, semi-natural or cultivated area.
Note: In addition to the use of regional specific crops and animal breeds, the conservation of regional or site specific additional flora and fauna is important for the in-situ conservation of agriculture. Until recently, such additional flora and fauna were often referred to as 'weeds' and 'bugs', indicating that they were not appreciated on cultivated lands. Only during the last decade, the ecological importance of additional herbs, which were called weeds before, beneficial insects or pest-predators became more widely understood, recognised and

appreciated (Altieri, 1992; Kiss *et al.*, 1997; Schotveld and Kloen, 1996; Van Mansvelt and Mulder, 1993).

Parameters for flora and fauna species' diversity:

1   Species diversity per bio-type and bio-tope
2   Targeted Plant Species Diversity (TPSD), Target Trees Index (TTI) and Target Shrubs Index (TSI)
3   Plant Species Diversity (PSD) and Plant Species Distributions (PSDN)

1   *Species diversity per bio-type and bio-tope*

For the assessment of a particular landscape on a particular scale, it is important to invite or at least consult experts of the local natural and semi-natural ecosystems. Mostly, different experts exist for different plant- or animal groups. Each group has its own particular purpose and significance, as perceived by different social, scientific and professional groups, which use often mixes of empathic and rational arguments.

This parameter can be used for the measurement of relevant standards for presence, abundance and guild structure of species diversity per bio-type and bio-tope or site.

The following aspects needs special attention:
- Many species of various plant and animal groups, endangered species included, are related to habitats which have been developed in pre-modernised agro-landscapes (Van Mansvelt and Mulder, 1993);
- Currently, most species' diversity tends to be found in agro-landscapes and farms, which are managed according to the concepts and standards of organic agriculture (Van Mansvelt and Mulder, 1993; Booy and Noorlander, 1992);
- Plant and animal species are largely inter-related to communities or guilds which are themselves related to specific habitat conditions for food- and shelter webs and hierarchies (Smeding, 1995; Thiele, 1997);
- All plant and animal species have their particular dependencies on and functions toward one or more other plant and/or animal species and the a-biotic environment like the soil-climatic conditions (Van Mansvelt and Mulder, 1993; Vereijken and Kloen, 1994).

Every perception of single species that focuses on one or more functions of the single species, looks independent and will easily overlook the multi-functionality and interrelationships that species have with their ecosystem partners. Although assessment of species and genetic diversity is necessary for the measurement of bio-diversity, the assessment itself can not be managed at any single species level. Therefore, the bio-tope diversity is a crucial parameter which have to be combined with the species diversity in order to assess bio-diversity.

The search for habitat specific species, which are usually higher (flowering) plants and vertebrate animals, is useful when it is carefully handled within the before mentioned context. Therefore, the identifi-cation of single selected taxa per type of land-use and region is still very useful.

Ecological importance of invertebrate species' diversity

Presence, abundance and diversity of effective predatory carabides, staphylinidis and spiders in natural as well as agro-ecosystems, considerably reduce the need for pesticides use (Booij and Noorlander, 1992). Agricultural fields can harbour a substantial number and diversity of predator species. More than 25% of the carabid species in North-west Europe occur in arable fields or managed grasslands (Thiele, 1997).) Mäder et al., 1995 stated that the earthworm bio-mass of well-managed arable soils should be > 120 g/m2.

Ecological importance of (well managed) farmlands for bird species' diversity

Farmland, covering 44% (141 million ha) of the EU area, is by far the biggest single main habitat type of the European countryside and the most important habitat type for threatened European birds (Birdlife International, 1997; Tucker and Heath, 1994). Feeding and breeding of 60% of all listed bird species with an unfavourable conservation status (UCS) depend partly or wholly on farmland. When farmland is appropriately managed, it has the highest number of regularly occurring species and the highest number and proportion of species with an UCS status (Birdlife International, 1997).

Farming practices affecting species diversity

Farming practices such as fertilising, grazing, moving, crop rotations and weed management affect species dominance and thus the diversity of different taxa, species and communities of flora and fauna species (Altieri, 1992; IEEP, 1994; Schotveld and Kloen, 1996; Smeding, 1994). See for instance the the example of Booij and Noorlander (1992) about the assessment of desirable invertebrate species diversity.Strategies for grazing and mowing can promote different plant and animal species, depending on timing and intensity of farming activities (Van Buel, 1996; Hagemeijer et al, 1996; Peeters et al., 1993).

Peeters *et al.* (1993) and Garcia (1992) showed that the highest diversity of species is achieved with a low intensity regime such as a one time hay cutting per year followed by late season grazing. More intensively rotational grazing together with winter grazing may still result at least in a high diversity of flora (Jenkis, 1987).

Assessment of desirable invertebrate species diversity
(regional and site specific)

Booij and Noorlander (1992) examined the population dynamics of carbides, predatory staphylinidis and linyphiid spiders. They showed that significant differences do exist between organic, integrated and conventional farming systems. The response of predators to the different ecosystems was measured for (1) predatory abundance, (2) species density and (3) guild structure or species-abundance pattern. Differences in abundance were found at crop level. Most carbides, staphylinids and spiders were most abundant in winter wheat and least abundant in carrots and onions. At system level, organic plots showed most abundance. See also Van Mansvelt and Mulder (1993), who quoted nine studies with similar results. The difference in species density did not differ at crop level and was less pronounced at system level. Guild structure in organic systems showed the most typical and complete species composition. The results of this research indicated that agro-ecosystems can be called 'diversified' for the above mentioned species if the following thresholds are reached.

Table 3.5 : Research results in number of animals per trap per year

Animal group/ crop-type	Winter wheat	Pea	Potato	Beet	Onion	Carrot
Carabids' Abundance	>700	>600	>400	>350	>300	>200
Staphylinids' abundance	>500	>250	>100	> 90	> 90	> 70
Spiders' abundance	>550	>300	>250	>250	>230	>150
Carabids' density	> 15	> 15	> 13	> 12	> 10	> 9
Staphylinids' Density	> 10	> 9	> 10	> 8	> 8	> 8

Source: Booij and Noorlander, 1992

More detailed information about the impact of grazing and mowing on the bio-diversity is available in Baars (1990), Etienne (1996), Fleury and Muller (1995) and Younie and Baars (1997).

Another example is the application of nutrients. Nutrient application strategies have controversial effects on species diversity, especially on grassland. In general, N and P application are supposed to promote greater productivity through the dominance of few vigorous plant species, which reduces the abundance of a large number species with low nutrient tolerating levels (IEEP, 1994). However, not only the presence and availability of N and P should be included in research, but also the mutual interaction and interaction of N and P with the soil organic mater have to be included in research in order to understand their overall effects. While, for instance Peeters et al. (1993) state that grasslands with many species are unlikely to survive when the level of soil extractable phosphate (EDTA-Acetate) is lower than 5 mg/100 g soil. Green (1990) states that only minor P effects on species diversity, as compared to N effects, do exist. Moreover, in the context of species diversity, the guild-links or webbing of species has also to be considered. For example, low input grasslands have a flora diversity, which also contributed to the diversity of fauna communities (IEEP, 1994). An application rate of 50 kg N/ha already reduces the number of earthworms, collembola and particularly myriapode populations (Edwards, 1984). For recent data see Younie and Baars (1997) and Ansaloni and De Roest (1997). See also the next criterion of bio-tope diversity.

Stocking rates (SR /SD) seem to affect the ground nesting bird populations. A SR of 2.4 cows/ha resulted in a loss of 40%-70% nests from lapwing, snipe and redshank, while a SR of 4.8 cows/ha resulted in a loss of 60-90% nests from these birds (O'Connor and Shrubb, 1986).

Also a diversified selection in crop rotation importantly affect the population of birds on arable land. For example, Wilson (1992) found that, contrary to the majority of broad-leaved crops such as oil-seed rape, legumes and roots, grass clover mixtures and under-sown cereals, which are both typical for organic farming, have a positive effect on birds breeding. The same favourable effects were found for diversified crop rotations, which are essential for organic agriculture, unlike rotations with mainly one or two crops.

As the application of agricultural activities has an impact on species diversity, it should be clear that organic and conventional agriculture have quite different impacts. Chamberlain et al. (1996) analysed and compared bird populations on 22 paired organic and conventional farms. They found significantly bigger and more bird populations on organic farms during the winter as well as during the breeding season than on conventional farms. A follow up study about butterflies at eight of these paired farms showed that significantly more butterflies were found on organic farms than on conventional farms. The number of non-pest butterflies were two times higher on the organic farms than on the conventional farms.

2   Targeted Plant Species Diversity (TPSD), Target Trees Index (TTI) and Target Shrubs Index (TSI)

The Targeted Plant Species Diversity (TSPD) indicates whether target plant species in the ecological infrastructure occur in space and time. The TPSD can be subdivided in the Targeted Trees Index (TTI) and the Targeted Shrubs Index (TSI). These three parameters differ per region and should be defined by local experts in relevant disciplines. For example, Kabourakis (1996) defined for the Cretan olive groves a TTI of tree species per 100m olive grove of >12 in hilly groves and > 8 in plain groves and a TSI of shrub species per 100m of olive grove of > 50 in hilly groves and > 70 in plain groves.

3   Plant target Species Diversity (PSD) and Plant target Species Distributions (PSDN)

Vereijken (1998) proposes to modify the TSPD parameter. He distinguished the Plant target Species Diversity (PSD) and the Plant target Species Distribution (PSDN). The PSD is the number of species with flowers, which are conspicuous through their colour and/or shape and attractive for fauna and tourists. The PSDN is the average number of target species per 100 m ecological infrastructure. For the Dutch province Flevoland, Vereijken proposes a PSD of > 50 per farm and a PSDN of > 20 per 100 m ecological infrastructure.

2.1.2   Sub-criterion: Bio-tope diversity

TARGET: To safeguard the ongoing sustainable development of a rich bio-tope diversity within a sustainable landscape setting. This refers to the un-cultivated, semi-cultivated and cultivated bio-topes of valuable and appreciated landscapes. The idea is to warrant per region the existence and, where relevant the ongoing development of characteristic plant and animal habitats as landscape elements, whether they are rare, threatened or still common. It is important for the sustainable development of rich bio-tope diversity to discuss explicitly the perceived functions and appreciation of characteristic plant and animal habitats. Such a discussion is necessary for good assessment per single bio-tope and per bio-tope diversity such as an ecosystem, web or guild, of how much bio-topes and bio-tope diversity are necessary and characteristic for the particular landscape area.

Note: For bio-tope conservation by agriculture it is important, in addition to the before-mentioned regional and site specific flora and fauna, to be aware of their mutual added value like forests and moors, wetlands and dry-lands, high moors and swamps. Until recently, bio-tope conservation was mainly part of nature conservation and supposed to be in conflict with agriculture. The idea of separating nature conservation from agriculture at large scale was to leave each other with its own interest. Nature in agricultural areas was largely referred to as wasteland, because it was not used for production and thus supposed not to be appreciated. Only during the last decade, the importance of wastelands for agro-ecological and landscape functions became more widely understood and appreciated (Baldock and Beaufoy, 1993; Van Mansvelt and Mulder, 1993; Van Mansvelt and Stobbelaar, 1997).

Parameters for the bio-diversity:

    1   Minimum standards for bio-topes per farm type

1   *Minimum standards for bio-topes per farm type*

Many studies suggest that a minimum of 3%-5% of the total farm area should be designated for on-farm nature conservation (Kabourakis, 1996; Smeding, 1994; Van Bol and Peeters, 1995; Vereijken, 1995). In addition to these (semi)-wild areas, productive farm areas like extensively managed grassland and orchards and set-aside fields may also contribute to on-farm nature conservation. To warrant the ecological infrastructure between the natural elements like corridors and stepping stones, the on farm nature conservation areas should not be concentrated on one spot only, but well distributed and scattered all over the farm. Production areas which lay between the on farm nature conservation areas should not be bigger than 8 ha (Smeding, 1995). Smeding recommends, for reasons of ecological diversity and ecosystems pest-control, a maximum field surface of 5 ha and a ratio between field width and field length of 0.8. For smaller fields this ratio can be 0.33. Schotman (1988) recommends 1,000 to 2,000 meters of linear woody elements per 25 ha as an optimum for field margins for some common bird species in the Netherlands such as the Curlew and the Partridge. Kabourakis (1996) recommends for the Mediterranean countries an ecological infrastructure area of 4% in hilly olive grove regions and 8% in plain olive grove regions. 15% of these ecological infrastructure areas should consist of non-linear elements and 85% of linear elements. In order to provide habitats for a variety of organisms, the vegetation of the on farm nature conservation areas should be at least 30 cm high during the winter and 80 cm high during the summer period (Smeding, 1995). From the above mentioned studies and recommendations, it seems to be clear that to warrant the sustainable development of regionally specific and characteristic valuable landscapes, the specific demands on presence and abundance of characteristic species and bio-topes should be elaborated and discussed by a multidisciplinary team of local experts. Generally, desired and specified ranges for the above mentioned aspects can not be given.

2.1.3   Sub-criterion: Ecosystems' diversity

TARGET: The target of the ecosystem's diversity is in line with the former target of sub-criterion 2.1.2 (bio-tope diversity) and with the main target of main criterion 2.1 (bio-diversity). This means that the target is to safeguard a sustainable development of the regional landscape biosphere diversity, within the context of a well-structured and well-cultivated regional and supra-regional network of ecosystems.

Parameters for ecosystems' diversity:

1   Minimum standards for types, numbers and size of ecosystems per landscape and region
2   Multifunctional landscape management
3   Regional specifications on presence (quality) and abundance (quantity)

1   *Minimum standards for types, numbers and size of ecosystems per landscape and region*

A joint study from the Institute for European Environmental Policy, the World-wide Fund for Nature and the Joint Nature Conservation Committee show that low intensity farming systems have a strategic value for nature conservation. This holds especially for species, which are mainly found on farm fields and for species, which are spread over a large area and cannot be protected within the confines of small nature reserves (IEEP, 1994).

2   *Multifunctional landscape management*

Other parameters for the presence and size of ecosystem types can be derived from the concept of multifunctional landscape management. As each function requires it's own specific set and number of bio-topes, multifunctional land-use will as such contribute to bio-tope diversity. To optimise bio-diversity in the context of the landscape as a whole, the coherence of the various bio-topes and natural elements has to be well structured. See also criterion 2.3 ecological coherence.

3   *Regional specifications on presence (quality)and abundance (quantity)*

A method to monitor bio-diversity or 'ecology-production' has been developed by Buys (1995). This method is called the Yardstick for bio-diversity. Although this method has been developed to monitor species diversity at farm level, it seems also useful for monitoring bio-diversity of bio-topes at landscape level. This method helps farmers, policy makers and traders to improve, support, promote and sell the farm itself and/or the bio-diversity of the region. An explanation of the Yardstick for bio-diversity is given in the next box.

Yardstick for Bio-diversity (Buys, 1995)

The monitoring method for bio-diversity consists of the following steps:

1. Selection of regionally relevant indicator species groups, which enable easy and reliable assessment and provide sufficient information about the actual and potential effect of farm management at on-farm bio-diversity.
2. Selection of individual species from the above selected groups, which presumably will be found in the agricultural area and respond on different farm management practises. Under Dutch conditions, this step resulted in a list of 199 species of vascular plants, 17 species of mammals, 77 species of nesting birds, 14 species of wintering birds, 7 species of amphibians, 2 species of reptiles and 26 species of butterflies.
3. Counting the selected species by using qualitative census methods such as plants, mammals and reptiles, nesting bird territories, maximum number and month of presence of non-nesting birds, number of egg batches or strings by the amphibians.
4. Rating the regional and national importance attached by society to the species, considering:
   - *Ecological importance* such as rarity and trends in population size. Ecological importance is valued by multiplying logarithmic calculations. This results in a rating value from 1 up to 100. Rare species with a negative abundance trend and great importance get a high score.
   - *Scenic value* such as plant height, colour of the flowers, flowering period, colour of the fruits. These values are added up and multiplied, resulting in values between 1 and 100. Tall plants, usually bushes, with colourful flowers and fruits, get a high scenic score.

The final score of the *yardstick for bio-diversity* is the product of the number of units resulting from step 3 and the rating score from step 4. Finally, the scores (3 x 4) should be divided by the acreage of the farm or the concerned bio-tope. Then the result is a relative farm and bio-tope. The yardstick seems a more reliable criterion in assessing ecology production than merely registering the presence of a species.

2.2   Main criterion: Ecological coherence

MAIN TARGET: As the main target of the column 2 (ecology) of Table 3.1 (Table of the Checklist) is an appreciated diversity of species, bio-topes and landscapes, this always presumes that diversity is found within a unified context. Otherwise unnatural sites like zoos, botanical gardens, flower-shops and pet-shops would be the ultimate examples of successful bio-diversity management. Here, the aimed diversity as a criterion for sustainability of landscapes, is an ecologically coherent diversity. Compliant to the before mentioned elaborated targets, the idea of this target is that each species can only figure and function within an eco-system of other flora and fauna species. Such a system relies on and contributes to the common environment of different species. See also the criterion of clean environment in column 1 of Table 3.1 (Table of the Checklist).

Note:Technically, crops and animals can be kept or produced in "hors sol" and "off-season" conditions by supplying them with an artificial environment including nutrients and waste-management. However, such an artificial environment does not produce the appreciated landscape, which society demands (Giorgis, 1995; Group of Bruges, 1996). In order to warrant the sustainable management of landscape, a keen awareness of the various connections and links of species and bio-topes with one another and their environments must be generated or at least encouraged.

This main criterion is subdivided into the following sub-criteria:

2.2.1   Vertical coherence: onsite
2.2.2   Horizontal coherence: in the landscape
2.2.3   Cyclical coherence: in time

2.2.1   Sub-criterion: Vertical coherence: on site

TARGET: The vertical coherence refers mainly to the site conditions of soil and water, but includes also slope and sun and wind exposure, within the local climate. Sun and wind exposure is co-determined by the regional vegetation and vice versa. The purpose of vertical coherence is that species and bio-topes should fit to the soil conditions they are growing on and deliver specific contributions from their eco-system to the environment they are cultivated in. See also the sub-criteria of column 1 (environment) in Table 3.1, viz.: fertile and resilient soil (sub-criterion 1.1.1), water quality (sub-criterion 1.1.2) and air quality (sub-criterion 1.1.3). As mentioned before, conservation of any particular development-stage or phase is not the main issue. The objective is to warrant a sustainable, ongoing development or evolution. However, making big costs through high external inputs, such as energy and other resources, in order to control the climate with physical technology like greenhouses, intensive husbandry and straightening rivers, exceeds the meaning of the vertical coherence target. Vertical or on-site coherence is conceived as an idea that facilitates the 'readability' of the landscape. In other words, when vertical coherence exists, the agro-silvi-pastural land-use expresses the local or site-specific conditions of the landscape. Then land-use gets to some extend a value as an indicator for the specific local conditions or local identity ('Zeigerwert'). See also column 5 (psychology) and column 6 (physiognomy/cultural geography) of Table 3.1 (Table of the Checklist).

Note:Whereas species-specificity is the focus point of species conservation or in German 'Art-gemaesse Anbau' or 'Zucht', here the soil- or site-specificity is the focus point or in German 'Boden' or 'Standort gerechte Anbau'. A demand for free ranged animal husbandry, which reflects consumers' awareness for site-specificity, is embedded in animal welfare organisations. For plant and animal species in nature conservation, a considerable discussion is dedicated to questions about autochthonous history of species. For instance what is the acceptability of import species form abroad? However, this discussion would by far exceed the limits of this study. Therefore, the discussion is only mentioned as an issue for debate for those who are involved in regional landscape and land-use planning. See also column 4 (sociology), column 5 (psychology) and column 6 (physiognomy/cultural geography) of Table 3.1 (Table of the Checklist). This target is presumed to follow up and to comply with the previously mentioned criteria.

Parameters for vertical coherence: onsite

1   Site specific indicator species
2   Site specific habitats and ecosystems

1   *Site specific indicator species*

In line with the famous work of Ellenberg (1988) ("Zeigerwerte der Gefaesspflanzen"), local experts can identify the flora and fauna species, which are characteristic for a specific site. (Agnew *et al.*, 1993; Novakova, 1997; Thompson *et al.*, 1993; Van der Maarel, 1993;). Such an identification activity can only be done per region and not in general. As the identity of a specific site is a product of its history, this parameter almost reach the parameters of the criteria in the sixth column of Table 3.1. For instance, when a site has a forest history, then the specific indicator species are others than when the site has an arable history. However, whether a forest or arable site, wet clay areas have different indicator species than dry limestone ones. The temporarily underestimated, but recently re-emerging knowledge of specific regional soil properties might play an important role to realise the importance of regional characteristics (Schraps and Schrey, 1997). See also the aspects of regional specific products in column 4 (economy) of Table 3.1. There seems also to be a link between the knowledge of 'old' plant species and 'old' animal species as they have been cultivated within groups. Examples are land-species, 'old' cow spaces and chicken species (Audiot, 1995; Gama *et al.*, 1997; Millar, 1997; Sambraus, 1994). Going back to the old varieties is not only important for the eternal conservation of species, but also for their big resource value toward a sustainable regional development.

The term *appropriately adapted* refers to the possibility of species to acclimatise, over a certain period of time, to new environments. The adaptive potential of species is an important issue in plant and animal breeding and related to the discussion on indigenous and exotic species (Oldeman, 1990). Further details about this subject exceed the subject of this study. However, it is mentioned here as an issue of careful consideration when talking about site-specificity.

As argued before, a difference can be made between two different approaches. The first approach is to push nature elements in an envisaged direction with the available technologies and so forcing to change the environment and/or the genetic structure of the relevant species. The second approach is the use of eco-physiological functions of species and bio-topes in order to gradually change their environment and eventually their genetic structure. The second approach merges aspects of evolution and succession, in such a way that elements of nature are invited to change according to their nature, instead of pushing them to adapt toward technological development (Lammerts *et al*, 1998;Temirbekova and Van Mansvelt, 1998 in press).

2 *Site specific habitats and ecosystems*

When talking about biological relationships, attention have to be paid to the habitats and ecosystems of species, which are their initially home in the short term and their substrate for subsequently changes in the long term. As the agro-silvi-pastural landscape per se is a product of culture, decisions on which parts of which habitats have to be left alone and what kind and size of habitats have to be created, are since long time in the hands of mankind. With respect to the previous mentioned targets and criteria, it should be clear that this parameter of site-specific habitats and ecosystems concerns the ecological functions of the various habitats towards each other. As all species have their specific functions within the different habitats, so do habitats have their functions in different landscapes. For instance, forests, meadows, hedges, ponds, streams, arable fields, slopes and meanders, all affect each another's micro and meso-climatic conditions like wind, temperature, shade, hydrology and soil structure. See also sub-criterion 2.2.2 (horizontal coherence) in Table 3.1. The more awareness exists about interactions and interdependency of habitats, the better 'either/or' opinions can be shifted to 'as-well-as' opinions. These 'as-well-as' opinions allow a creative discussion on the locations and types of habitats in order to warrant which landscape functions for which habitats. The debate about multi-functionality succeeds the debate about the single functionality which deals only about one, most important, function.

The issue of this parameter is to make sure that the various habitats fit to and show the local and site-specific conditions, instead of being (artificially) plugged in at the site. The plug-in approach has been quite characteristic for modern agriculture and forestry: on which site grow the most financially profitable crops on the maximum available areas wherever possible and using considerable amounts of external inputs and applying considerable land reclamation actions. This process of modern agriculture and forestry resulted into a striking homogenisation of European landscapes at the cost of regional landscape diversity and site-specific habitats and not to forget soil erosion and water and air pollution. See also the environmental criteria in the first column of Table 3.1.

2.2.2   Sub-criterion: Horizontal coherence: in the landscape

TARGET: Here the issue is to organise the landscape in such a way that various types of land-use, and so the various cultivated habitats developed and/or protected, do show synergy as indicated in the previous criteria. The ecology of habitats within a landscape refers to an up scaling of the species' ecology within a habitat. Gradients from natural ecosystems, like forest boundaries and riverbanks, are well known for e.g. drought, salinity, altitude and temperature. These gradients merge the site-specificity with the landscape coherence in an obviously natural way. For cultivated landscapes the challenge rises to find and establish a cultivated form of horizontal coherence, which is possibly upgraded or at least not degraded. Such a horizontal coherence exists not always in the form of gradients, but often in some form of mosaic. Then, such mosaics may follow geo-morphological structures and fit into the landscape, or may deny them and look abstract and/or not fitted. In the ecological framework, the size of the mosaic parts is important either to facilitate or to break connectivity of species and food or prey interactions.

Note: This column 2 (ecology) of the Table of the Checklist (Table 3.1) still largely relies on the natural science approach in which the horizontal connectivity refers to ecological connectivity or webbing, with some reference to labour intensity through farm design and labour efficiency. In columns 5 (psychology) and column 6 (physiognomy/cultural geography) of Table 3.1 the visual coherence will be discussed from an aesthetic point of view.

Parameters for the horizontal coherence: in the landscape

    1   Species coherence
    2   Habitat and eco-system coherence

1   *Species coherence*

How spatial patterns of landscape elements like arable fields, woodland, hedges, ponds, etc. facilitate the mobility of organisms, including pests and pathogens, and influence the cyclical mobility of temperature, air, water and chemical compounds has been described by Ryszkowski and Kedziora (1995).

Barret (1992) indicated several assessment indices for the various levels of diversity:
- Alpha diversity indices within the habitat;
- Beta diversity indices between habitats;
- Gamma diversity indices at landscape level, which describe large-scale patterns, processes and phenomena.

Shannon Weiner Index (Barret *et al.*, 1990)

Barret *et al.* (1990) used the Shannon-Weiner index to measure the decline in crop diversity in Ohio for the period 1940-1982. They found that crop diversity decreased from 0.80 in 1940 to 0.60 in 1982. This decline originated mainly from the elimination of small grains and hay out of the crop rotation and was correlated to the elimination of winter habitats for beneficial insects and the enhanced movement of pests between crop fields.

---

2 Habitat and ecosystem coherence

Based on a long-term research in West Poland, Ryszkowski (1995) suggested that a diversified mosaic landscape comprising of arable fields interspersed with woodland, shelter-belts, hedgerows, (riparian) meadows, swamps, ponds and ditches enhances water storage, controls groundwater chemistry and helps to maintain biological diversity.

At farm level, Smeding (1995) distinguished wet and woody elements in the on-farm ecological infrastructure. Since both, wet and woody, elements provide habitat for a number of different predatory organisms with different action radiuses, it is essential that both types of elements are properly distributed and connected over the entire farm area. Krekels (1994) as well as Smeding (1995) proposed a minimum distance of 400 m between each wet element like ponds and channels, and 100 m between each woody element like a group or line of trees and/or shrubs. The water level differences between different channels, or between part of channels, should not be more than 10 cm, to allow migration of fish. Similarly, the profile or diameter of tubes connecting water bodies should not be narrower than 50 cm, to allow migration of water mammals (Boer, 1993; Smeding, 1995). Single trees which stand on their own, provide resting and nesting places for many animal species, but are also important references or orientation points for animals and humans (Smeding, 1995). Field margins of 1-2 m wide should be established along all woodland edges, hedges, bug banks and ponds, to provide a buffer between the wildlife habitat and agricultural activities. Manure and pesticides should be applied at a minimal distance of 3 m away from hedge bottoms, bug banks and the margins of the permanent and intermittent ponds (FWAG, 1995). Based on studies from Van Heudsen *et al.* (1994), Schmitz (1993) and Smeding (1995) a group of trees and shrubs of 10x3 m as a minimum woody area element is recommended. Ponds should have a 50 to 200 m2 water surface and at least a depth of 1 m as a need for regular cleaning and to prevent land growing. The slopes of watersides should be 20% on the South and 10% in the North exposed slopes. A mix of permanent ponds and temporary pools provides a variety of aquatic habitats, which is important for a range of invertebrates like snails, beetles, caddis flies, damselflies and other flies.

Next to the above mentioned ecological infrastructure elements, some special habitats built from woody blocks, stone and straw heaps, chopped wood, piles of fallen tree branches, etc. should be established at a farm to warrant optimal mammalian and other species diversity (FWAG, 1995; Smeding, 1995). Stonewalls and stone-heaps, well known in the UK and Mediterranean landscape, provide moist and cool shelter in summer and dry warm shelter in winter. Since most of the bird species require a quiet place for breeding, Hund (1994) and Smeding (1995) recommend that nature conservation sites and other on-farm ecological infrastructure elements should at least stay about 200 m away from noisy places such as public and farm roads, stables, machinery shelter, etc. Insects, amphibians and reptiles require 'sun-bathing' to reach their active temperature. This means that considerable parts of nature conservation areas should be exposed to direct sunlight for about 12 to 15 hours (Smeding, 1995). However, considerable shadow areas are important for other species and a shelter for sunbathers. So here again, the issue is not an 'either or' but 'as well as'. Appropriate diversity counts!

At national level, the Dutch government developed, ten years ago, a national nature policy plan to outline where, how and what parts of nature should be conserved, rehabilitated, developed and improved, and also within the period of time that the plan should be accomplished. The purpose of the plan was to build up a spatially stable national ecological network consisting of:

1   Core areas of usually more than 500 ha with a significant ecological value and national and/or international levels. Examples: forests, estates, stream valleys, dunes and large lakes.
2   Nature development areas with realistic prospects for the development of significant ecological values at national and/or international levels. Examples: nutrient poor or wet grasslands, marshland and marshy woodland.
3   Ecological corridors, which are areas or landscape structures, that contribute to restore the possibilities for migration within and between core areas.

The sustainability of the network is supported by a buffer policy to eliminate or minimise external influences on the network and a curative environmental policy to support management measures that enable ecosystems to recover from the acidifying and euthropicating effects of atmospheric deposition (LNV, 1990).

2.2.3   Sub-criterion: Cyclical coherence: in time

TARGET: The development of all natural organisms is facilitated by a complex of interrelated processes with each having their own specific time-span as an intrinsic quality of their existence. Those interrelated processes have various phases like emergence, growth, culmination, decline, breakdown and dormancy. The point is that in modern land-use systems or agro-sylvi-pastural production systems lob-sided attention has been given to the fast growing processes of the younger phases of life. This holds for plant as well as for animal production, where the purpose is to strive for the most rapid harvest of largest specimen with highest return-on-investment. Wherever one looks, in all cases the youth-phase is overexploited at the cost of the flowering and ripening phases (Baars and Bloksma, 1995; Koepf et al., 1996; Van Mansvelt, 1981). Examples of such cases can be found in the manure and climatic control in plant production like hors sol greenhouse hydro-cultures, selection of fast-growing species in forestry and grassland like spruce, eucalyptus and English ryegrass, early milking and dairy breeds or small early fruit trees. So, in order to warrant a full diversity of nature, also a diversity of life-phases in each of the ecosystem species and landscape bio-topes should be targeted. For instance, young and old grasslands, forests, trees, animals in herds, animal species in landscapes, etc.
Note: Next to the issue of presence of various phases of the lifecycle, is the issue of the lifecycle coherence with the seasons. Here again, modern agriculture has managed to surpass many of the limits of natural cycles through biochemistry and/or physical technology. If natural biosphere has been evolved as an

intrinsic part of the solar system, where the sun is not only the main source of energy but also the major pacemaker or biological clock, then it seems to be relevant for sustainable land-use to fit farm management practices into the seasons. This means fully respecting and following the seasons instead of counteracting the seasons at the cost of considerable amounts of non-renewable energy. Farm management, which respects and follows the seasons, will be clearly visible within the landscape as it expresses the seasonal qualities of the landscape. Of course, this target of seasonal coherence should not be over-stressed. Being aware of the value of cyclical coherence will affect decision-making. As this issue has not been very much studied yet, creating awareness on its possible value is the main purpose at the moment.

Parameters for cyclical coherence: in time

1   Full lifecycles of species and systems
2   Seasons compliant management: availability of nectar for 'flower-insects'
3   Seasons compliant management: timely differentiated hedge and woodland management
4   Seasons compliant management: timely management of water-bodies
5   Seasons compliant management: timely management of permanent pastures

1   Full lifecycles of species and systems

This parameter focuses on the presence of the lifecycle phases of species and bio-topes within a landscape. For example, in the 'wild' margins of arable fields can herbs pass their whole lifecycle according to their embedded seasonal pattern (El Titi, 1992). Organic grasslands in moderate climates show dandelions in spring (Taraxacum) and (Leontodon) in the late summer.

> **Example about organic farming and the landscape**
>
> A recent study, commissioned by the UK Countryside Commission on the effect of organic farming on the landscape (ENTEC, 1995) found that organic farming does have a positive impact on the quality of landscape. In comparison with the conventional counterparts, organic farms proved to be richer in diversity like young and recent hedgerow trees of various types, woodland and weeds at the crop fields. However, this difference was more pronounced in the lowlands with arable and mixed farming, than in the highlands with husbandry centred farming and in horticulture with vegetable centred holdings.

Under non-organic farm management with intensive use of NPK, dandelion appears in both seasons, spring and summer, if any flowers appears at all. Forest management has indicated that the presence of decaying trees is valuable for the forest ecosystem development (Kouki, 1994; Oldeman, 1990; Tyrell and Crow, 1994).

A recent Dutch study about landscape production of organic and non-organic farming systems shows that landscape of organic agriculture clearly tends to have more seasonal coherence than the landscape of conventional agriculture (Hendriks *et al.*, 1997). The full lifecycle concept also applies for animal husbandry and wild animals (Bockemühl, 1992; Kiley Worthington, 1993; Koepf *et al.*, 1996).

2   *Seasons compliant management: full availability of nectar for 'flower-insects'*

For the availability of sufficient nectar for 'flower-insects' throughout the season, the ecological infrastructure should contain a certain number and diversity of flowering and thus nectar producing species (Smeding, 1995). Attention should also be paid to the timing of blossoming, which means that nectar is available from early March till late October. Obviously, supply of nectar will be regular at farms with rich pastures, late-mowing meadows, abundance of weeds at the fields or field-margins and species rich woody elements like trees, hedgerows, etc.

3   *Seasons compliant management: timely differentiated hedge and woodland management*

Woodland edges or forest skirts are important nesting habitats for singing birds such as the song thrush and blackbird. Woodland edges that form field boundaries should be cut, depending on the re-growth rate, at a 5-10 year rotation, to prevent over-hanging and overshadowing of the herbaceous flora in the hedge bottom. Prevention of overhanging and overshadowing refers to the importance of both shade and sunlight and advocates a variation in hedge cutting. Thus to prevent a monotony and biased habitat formation. Sufficient diversity of species in hedges and selective cutting is also to warrant the availability of various bird-nesting habitats and bio-topes for mammals and invertebrates (FWAG, 1995). Where necessary, stock fences, distanced 1.5 to 2 m apart, should be placed around newly planted hedges to protect them against damage from cattle, rabbits or deer (FWAG, 1995). Few hedges should have free growth and create some hedge junctions. Surveys of birds that have their nests in hedges, indicate that the most favoured territories are those which include at least one hedge junction (FWAG, 1995). In order to perform full ecological and multipurpose function of hedges, they should be at least 3 m wide and 4 m high. Moreover, they should be allowed to flower and fruit at least every 2-5 years with a cutting frequency adapted to the species present (Stopes *et al.*, 1995).

4   *Seasons compliant management: timely management of water-bodies*

Ponds, ditches, canals and other water-bodies should be best cleaned on a 3-5 years rotation. The best period to clean water-bodies is between late October and late January, because then it is not the breeding season. For canals and other watercourses it is recommended to clean one bank-side at a time, to warrant the presence of site specific bio-topes. However, local and/or national regulations for water-body cleaning and maintenance have to be respected. Perhaps, it will be possible to re-open a dialogue about favourable strategies. Nutrient rich silt should be taken away from species rich banks, hedge bottoms or pastures, to prevent invasion of aggressive plant growing like nettles, cow parsley and thistles (FWAG, 1995). Highly fluctuating and complete dried-out water levels are undesirable for most water-bodies and should therefore be avoided (Smeding, 1995; FWAG, 1995). In the U.K., grants have been provided for good management of ponds, reed-beds, scrapes and water meadows. In Germany, restoration of meandering river-beds has been supported. And in the Netherlands, grants have been provided to restore old puddles (Beissmann, 1997; FWAG, 1995).

5   *Seasons compliant management: timely management of permanent pastures*

Pasture management should aim at the improvement of plant diversity of the grassland and reduction of scrub encroachment. This can be reached by (1) mowing after the seeding period of herbs and grasses and after ground-nesting-birds have fledged their young ones, (2) clearing encroaching shrub and (3) grazing animals with sheep after cattle (DLG, 1997; FWAG, 1995). To give options for wildlife to escape, hay and silage should be cut inwards (centrifugal) or in sections. As an incentive for farmers to manage grassland in a way that favours breeding of the meadow birds, the Dutch government has introduced nature management agreements with farmers (DLG, 1997). Farmers who want to receive the money from the management agreement, have to maintain a high water level, us limited amounts of fertilisers and preferably organic manure, postpone mowing and cattle grazing during the breeding season and ban the use of pesticides. The management agreements include a management fee and a compensation for the decline in income. However, it took 6 years before the first agreements were signed (DLG, 1997).

2.3   Main criterion: Eco- regulation

MAIN TARGET: In order to warrant and develop the species and habitat diversity, which makes a landscape ecologically interactive and attractive, the agro-sylvi-pastural management together with the management of the landscape infrastructure, should focus on pest prevention instead of fighting pests. Strategies for chemical fighting against pests, tend to systematically generate resistance of pests to pesticides instead of systematically stamping out pests (El Titi, 1992; Kenmore, 1991, 1997 and Kenmore *et al.*, 1997). Research in integrated pest management and organic agriculture show that creating crop growing conditions that prevent favourable pest conditions and warrant the presence of pest predators does certainly pays off (Lampkin, 1990; Vereijken, 1996a, 1996b). This obviously demand quite professional management, which can be acquired by training farmers and by exchanging farmer-to-farmer experiences.

Note:The first two columns of Table 3.1 (Table of the Checklist), environment and ecology, describe many aspects of ecologically sound management requirements. Good management may strongly reduce the occurrence of pests (Altieri, 1995; Van Mansvelt and Mulder, 1993; Vereijken, 1996a, 1996b).

Parameters for eco-regulation

1   Degree of pest and disease occurrence
2   Pest predator presence

1   *Degree of pest and disease occurrence*

The degree of pest and disease occurrence indicates not only the presence of diseases or pests, but also the conditions, which favour the presence of diseases and pests. This qualitative parameter with an indicative value may range from health or auto-immune-system or resistance, through acquired resistance to sufficient pest-predator bio-tope presence and absence of stress factors, which affect the health of crops and cattle (Edwards-Jones, 1997; Massales *et al.*, 1997; Michelakis, 1997).

2   *Pest predator presence*

As an indication for the presence of pest predators, one could check for bio-topes with minimal 2 predator species per relevant crop or cattle and pest, that should be present within the land-use system (Altieri, 1995; El Titi, 1992; Schotveld and Kloen, 1996; Vereijken, 1996a, 1996b).

2.4   Main criterion: Animal welfare conditions

MAIN TARGET: Many aspects of animal welfare have been mentioned before. However, mostly indirect and often focused on wildlife. Here, the focus is on animal husbandry conditions and restricted to some major considerations, which have to be taken into consideration when working towards sustainable rural and agro-sylvi-pastural landscape management (Spinelli and Baldini, 1993; Waterson, 1994). Most important aspects in general for cattle and husbandry with respect to sustainable landscape management are:

- Cattle and husbandry should fit and respect the carrying capacity of the regional ecosystem (see column 1 of Table 3.1 (environment));
- Cattle and husbandry should be instrumental to the landscape management. Intensive, 'hors-sol' and outdoor livestock production systems do not fit into this concept.

However, at the moment, society may largely prefer other concepts and priorities, such as cheap meat and high returns on investments. Examples of landscape management with livestock are dairy farms in the mountains of Switzerland, Austria, and Norway and nature conservation areas with sheep and 'wild' horses or ruminants (Kiley Worthington, 1993; Mearns, 1996).

Note:Considerable literature about animal welfare conditions is available (Baars en Buitink, 1995; Boehnke, 1997; Dousek, 1995; Ekesbo, 1992; Mathes, 1995;

Ratheiser, 1996; Rist *et al*, 1992; Sundrum, 1993). Parameters to be specified, per species and per region, with local experts.

Parameters for animal welfare conditions

1.1   Space for natural behaviour
1.2   Shelter against adverse weather
1.3   Preventive health care

1   *Space for natural behaviour*

In general, animal welfare conditions have to be in full compliance with EU and NGO standards. At the moment minimum requirements focus largely on space for natural behaviour (Von Borell, 1996).

2   *Shelter against adverse weather*

Spatial and temporal coherence as mentioned in the sub-criteria 2.2.1 (vertical coherence) and 2.2.3 (cyclical coherence), can be very well implemented to provide all livestock with sufficient shelter against adverse weather like sun (radiation and heat), wind, rain and cold. The same coherence can also be implemented to provide livestock with species specific food. This means that the type of food complies with the eco-functionality or resource efficiency of animal species (Kiley Worthington, 1993; Rist *et al.*, 1992; Van Mansvelt and Mulder, 1993).

3   *Preventive health care*

In line with the before mentioned main criterion 2.3 (eco-regulation) for preventive pest and disease management, this parameter goes for preventive healthcare. Low stress conditions and a healthy environment allow for a minimal use of antibiotics and hormones. Especially in organic agriculture, herbal livestock pharmacy becomes more important in combination with the low stress livestock conditions as preventive healthcare. Here again it is important to stress that pest prevention here means warranting such husbandry conditions, of housing, feeding and breeding that diseases do only minimally occur. It does certainly not mean preventive medication with pesticides (Baars and Buitink, 1995; Boehnke, 1997; Rist *et al*, 1992).

3.2.2   CRITERIA FOR THE SOCIAL REALM: ECONOMY AND SOCIOLOGY

J.D. van Mansvelt and M.J. van der Lubbe

Introduction

In the development of sustainable rural (agro-sylvi-pastural) landscapes, norms, attitudes and processes of the socio-economic sphere do play a most crucial role. All decisions made are based on some kind of prioritising, in whatever group(s) in charge for whatever part of the landscape. Although the landscape is usually seen as an object of either natural or anthropological sciences, underlying all decisions that earlier or later sort visible effects, deliberations on values, feasibility, profits, and interests have been weighed and traded out in some sort of transparency. People take decisions on all kind of aspects with all kinds of arguments and these decisions take place within social structures. In the end, these decisions are reflected in the landscape. As already indicated in the chapter 2 "Research methods", here the position is taken that it is the social realm wherein the perceived values of nature and culture are weighed against one another. Thus, we see this major realm as an intermediate between both other major realms viz. the (a)biotic realm and the psycho-cultural realm.

Here, the social realm, representing the qualities of the social environment, has been subdivided into economics (goods, money and services) see column 3 of Table 3.1 and sociology (power and access to responsibility for decision-making) see column 4 of Table 3.1. Criteria for the quality of the landscape's socio-economic environment are included in the checklist of parameters for a sustainable landscape development, because appropriate flows of money and compatible structures of decision making are required to allow for and warrant management on all relevant levels to meet the quality demands of the (a)biotic and the cultural realms. Thus the economic and social criteria represent the trade-out area between the human (and society's) physical survival and it's ethical survival (individual development). The way people and societies decide to spend their money and to participate in socio-political activities reflect their *empathic coherence* with the sources of the purchased product or service and or the ideals pursued in the actions they support. From this point of view, buying and political decision-making are both phenomena of implemented *sympathy* or *antipathy*, or in other words phenomena of implemented *engagement* or *alienation*. This can be found clearly reflected in the original market status, where the acts of meeting (socialising) were at least as important as those of the trading (economising), with the bargaining as an expression of the quality of personal meeting.

Although it is now widely perceived that these days most decisions are made on economic grounds, the identity of actual beneficiaries of those economic considerations may not be always very clear to the public. For example, few actors are fully aware that economic "laws" are the reflections of historical and regional habits, attitudes and appreciation of societies. Thus they are open for modification by societies if they want to. For example, perceiving the farmer's income as necessarily based on the sales of their food, feed and fibre production, and the landscape as an issue of public services, is a socio-political choice with considerable impact. Perceiving farmers as the major landscape managers, with the food and fibre production as main products but landscape as an inseparable side product to be fairly remunerated, is another one. Calculations on the economic effects of such decisions fully depends on the factors included in the calculations and on the number of factors which are kept fixed in constraint to changes ("business as usual"), versus factors allowed to shift in compliance with the new policy. The dilemma is to generate sufficient changes to reach the targets set, without however changing too much of vested interests' positions and prospective. Opting for win-win solutions, to be reached within an acceptable time-span, seems the only way-out.

For the development of a checklist for sustainable landscape management, however, hard figures about norms and standards for the economic and social parameters warranting a sustainable management of the landscape, are hard to find. Nevertheless, the importance of the parameters as presented, soft as they may be, is that by including them explicitly in the discussions, the freedom to make clear and fully conscious choices can be increased. Here we presume that the more trade-offs, together with the inevitable or optional links between the trade-offs, are known, the less surprises or unforeseen side effects will occur, once the decisions are taken and implemented.

In this study, equity in the sharing of the limited resources and equality in the participation to decision-making – with compliant sharing of the responsibility - are seen as the leading objectives in the social science realm at large.

Society's ongoing specialisation and spatial concentration of production over the past decades, as part of world wide industrialisation and urbanisation, has lead to structural changes in agriculture as well, going along with increasing dismissal of labour and increasing inputs of fertilisers, plant protection products and energy (mechanisation, processing and transportation). Moreover, agricultural regions have developed in a very different way, each according to its particular location, market access, and the farmers' ability to adjust their holdings to changing economic conditions (Hagedorn, 1996).

For society as a whole, the process of specialisation and spatial concentration led to an enormous increase of production and consumption, together with increased flows of goods, services and finances. In the rich countries this even led to overproduction and over-consumption, in particular of animal proteins. It also went along with increasing power and wealth for less and increasing dependency for the others. In the biosphere realm it led to an increased waste of resources and an increased production of refuse. It also led to the removal of wildlife habitat and landscape features. All these changes led to a painfully perceived reduction in environmental quality and the loss of rural area's multiple values for other uses (Baldock and

Beaufoy, 1993; Dabbert, 1997; Hodge, 1991; Lundgren and Friemel, 1994; Van Mansvelt and Mulder 1993; Wascher, 1997).

To overcome the negative (side) effects of this development, a well-balanced coherence between society's vested and future interests seems an objective worth striving for. This statement holds for society's industry as a whole, as well as for agriculture (land-use and landscape management) as one of its specific sectors. At the moment, agribusiness has a lot of power in the whole chain of food, feed and fibre production, ranging from farm (primary production) to household (final consumption). Agribusiness at large still tends to demand for uniform standards: same units, same quality, same product types (homogenisation and mono-cultures). Such a specialised economy tends to lose the benefits of diversity (Elzakker et al., 1992). And so does a specialised agricultural sector, wherein farmers get fully depend on the demands of agribusiness (not necessarily consumers'), and thus restricting their farming to the production of large quantities of a limited number of products (raw materials for industrial upgrading).

Kieft and Verberne (1998) evaluated the Dutch policy for rural development. They applied the concept of the reorientation of rural areas, meaning the optimisation of the rural area's functions according to social demands changing from only economic functions towards the inclusion of social, cultural and ecological functions (like outdoor recreation, landscape appreciation and enjoying semi-wildlife). To get implemented such a reorientation requires other perceptions about their roles from producers, consumers and policy makers. Functions for a well managed rural landscape are: carbon fixation, counteracting atmospheric pollution and climate change, biological clearing and storage of water, reduction of avalanches and soil erosion, resources for education and science, the values of cultural heritage, and the presence of a wide range of sensorial information.

The valuable and appreciated rural landscapes are in large part the result of private decisions made by farmers concerning the management of their farm activities or resulting from farmer organisations lobbying. The individual farmers' decisions affecting the landscape are taken within a common culture, giving incentives as well as setting limits to specific practices. However, lobbying for governmental ruling on landscape and environmental issues, usually takes place in a broader –industrial economy driven- context, while its results cover only selected factors, benefiting few subjects (Bennet et al., 1990). If today's private producers of valuable rural landscape do not receive any contribution to the costs of landscape production, they will be reluctant or unwilling to invest in landscape production or maintenance. For this reason, the uses of rural land, apart from producing food and fibre, tend to be under-supplied. It is therefore that some institutions, other than the conventional market, are necessary to stimulate the provision of valuable rural landscape (Latacz-Lohman, 1997). Examples of such institutions are private organisations, nature conservation, law and regulations, management agreements, environmental co-operatives, etc.

Within the decision making processes at individual, community, regional, national, or international level, interests are weighted between the short term interests of private parties and relatively small groups at one side, and the long term communal interests of the general public at the other. The own interests as perceived by those participating, can not but play an important role in this weighing process. For instance, at first sight it seems in anybody's own interest to get food at the lowest available prices. However, on second sight it may also be everybody's own interest to make sure that the food producers do not degrade the (their) land and the (common) environment, and therefore should earn enough money to survive when farming in a sustainable way. An example at international level is the interest of the rich countries to remit the debts of poor countries in such a way that the inhabitants of poor countries are able to buy export products from the rich countries' industry. Export of staple foods must however be considered in a very critical way, as it can easily misbalance the importing countries' agriculture and subsequently its rural society and landscape.

Quality of the social environment: Criteria for the social realm

Economy	Sociology	
Flows of finances and services	Participation procedures	
3.1 Good farming should pay-off	4.1	Well-being in the area
3.2 Greening the economy	4.2	Permanent education
3.3 Regional autonomy	4.3	Access to participation
	4.3.1	Farmers' involvement in activities outside the farm
	4.3.2	Outsiders' involvement in farm activities
	4.4	Accessibility of the landscape

3   ECONOMY (Column 3)

The economic problem is how to use limited resources to best meet the needs of society – needs for intellectual stimulation and personal development, for leisure, for consumer goods etc. (Lundgren and Friemel, 1994). The simplest way and the most common solution to this problem is to rely on markets – on the price mechanism. The free market system is presumed to reproduce the various partial interests, that together form the general public's interest, which then includes one's own and everyone else's interest (empathic coherence). Total demand then expresses the consumers' willingness to pay for various quantities of a given set of available goods or services of a certain quality. The free market's (price-) mechanism system is ideally based on the following conditions: (1) many suppliers; (2) many demanders; (3) homogenous products; (4) a transparent market; and (5) a free entry into the market. However, especially conditions 3, 4 and 5 are far from being met. Moreover, De Groot and Wagenaar (1992) summarises the free-market failures as follows: (1) exclusion of external effects; (2) exclusion of 'free' works of nature; (3) existence of different market places; (4) differences in perception; and (5) consumers' surplus. In practice, the market at large is controlled by a few international blocs, which have a decisive influence on the fixing of prices and trading conditions. Thus, the participation in setting market prices is restricted to a rather limited group of organisations with the main economic and therefore also considerable political power. World-wide market leading companies (for example in oil, chemicals, shoes, drinks) are dealing with a much "freer" market than national organisations producing local fruit juices or whatever other products in developing countries. The leading companies have the money and power to create the infrastructure needed to push their products in the world-wide market. Parallel it could be argued that the market for conventional agricultural products is freer than that for organic products.

To make economics compatible with the demands of the physical survival for society and ecology it obviously should be redirected, focusing on the identification and (re)allocation of costs and benefits (Bojo, 1990; Constanza, 1991; Daly et al., 1990; Hanley, 1991; Panayotou, 1995; Tellarini et al., 1996). See also the environmental and ecological criteria in column 1 and column 2 of Table 3.1 (Table of the Checklist). This regards the various scales from farms, through local communities, villages and regions up to national and international economics. Therein, a central point is the identification and re-internalisation of the environmental costs that previously have been externalised. This complies with such conceptual principles as (1) the non-polluter benefits (NPB) and (2) the polluter pays principle (PPP). This regards the whole hierarchy of relevant levels, ranging from farms through local and regional communities up to the national and international ones.

However, economy as mentioned, is strongly connected with juridical regulations affecting the allocation of costs. For example the distinction between external costs and external benefits turns on the juridical issue of whether a farmer has a right to undertake a particular action or to cause a particular effect (Hodge, 1991). If for example a farmer or whatever other producer does not have the right to destroy a valuable habitat on a private property, its destruction is regarded as the production of an external cost. On the other hand, if he <u>does</u> have this right, his choice <u>not to destroy</u> is regarded as the production of an external benefit. So the payment of income compensation to farmers that do not contaminate the ground or surface water implies that they do have the right to allow pollutants to leach from their land. This then means that the reduction of such emissions presents an external benefit (Hodge, 1991). Such decisions, with their major effects on economy, draw heavily on society's perception of justice in private property affairs. In general, a valuable and appreciated landscape is regarded as an external benefit, and thus as a public good, for which consumption is non-exclusive and non-depleting. Landscape is then perceived as an immaterial product, like music or painting. "Non-excludability" means that once the product is available for consumption by one consumer, it is not possible to prevent other people from consuming it as well. The second term, "non-depleting" indicates that the consumption of a good like the landscape by one person does not reduce its availability for consumption by anyone else, at least in the absence of any deteriorating effects of congestion or over-population by massive effects (Baumol and Oates, 1988). But the non-depleting character of certain goods is not unlimited, as for example over-population and a massive visiting of a landscape will destroy it.

Three main criteria have been distinguished with respect to the flows of finances and services, viz.:

3.1   Good farming should pay-off
3.2   Greening the economy
3.3   Regional autonomy

3.1   Main criterion: Good farming should pay-off

<u>MAIN TARGET</u>: A prerequisite for rural landscape management is the survival of farmers and communities in rural areas. This means that farmers' subsistence and thus farming systems' subsistence should be warranted. Therefore, good farming should pay-off, to make sure that the good farmers remain or move to the rural areas in need of good agro-landscape management.

However, what is meant with 'good' and how should that be paid? Environmentally sustainable farmers or farm co-operatives, in one way or another, should receive a regionally acceptable income per labour unit. Conventional concepts have stimulated farmers to produce as much as possible, which resulted in overproduction, low prices and environmental costs. For instance, Dutch farmers have been stimulated to produce the most profitable crop at the costs of the environment and employment. Daly *et al.* (1990) and Hoogendijk (1993) pointed out in their theory that such concepts can only hold if society is willing to accept the environmental costs and the costs of unemployment. This means that if society wants to have cheap food, it should accept environmental costs and unemployment. In former days, society paid a relatively higher price for milk, compared with current days. This higher price included a contribution to the labour and production circumstances such as soil fertility and the landscape, which viewed in retrospect to be valuable. A current example where prices include labour and production circumstances are certified organic (many labels) and bio-dynamic products (Demeter).

In order to get access to the financial status of farm households and co-operatives, good transparent bookkeeping is required as a prior condition for paying good farming and to find out about the actual costs of good farming. Appropriate spread-sheet instruction support programs could be helpful and should be made available to farmers and co-operatives in such away that a farmer knows how much income he earns and how he earns this income. For instance, it should be possible to derive the level of market orientation (cash-cropping) and subsistence farming from such spread-sheet bookkeeping. The links between capital involvement and responsibility for final decision-making have to be considered. Moreover, attention should be paid to the types of subsidies available and their effects on different types of farming.

Another pre-requisite to pay for good farming, is transparency in the management, tenure-ship, and ownership situation of land to allow assessment of the decision making and responsibilities. In general, it is expected that people will make proper use of anything they have a property right on. For instance, it is often stated that depletion of communal lands is caused by a lack of property rights: the tragedy of the commons. However, it is not only purely property rights that will cause proper and sustainable use of land, but also the responsibility of land-users towards society and towards the future land users (successors). Farmers' who have private ownership on their land, can do with it what they want till a certain limit. The limit is set by those people, which get affected by the negative external effects from non-proper and unsustainable land use. Mostly, limits are set by direct stakeholders or by society as a whole. In case of sustainable landscape and nature management, there are various organisations, including farmers, who claim to offer 'best' management. There are private landowners, which mention that it is not only the economic value of land, but also the emotional value, which stimulates them to take care of their land (De Hen and Van Leeuwen, 1997).

Parameters for good farming should pay-off:

1 Total net farm income
2 Total farm family income
3 Return on labour
4 Farm's market orientation
5 Financial autonomy

1 *Total net farm income*

This parameter is one of the monetary figures that give insight in the farm management. It shows how much cash income a farmer receives from his on-farm activities. A certain level of farm income is necessary for a farmer to continue his farm activities. A farmer should know which of his environmentally sound farm activities are profitable. Since the survival of good farm practices and thus of farmers is necessary for sustainable landscape production, a certain level of farm income is needed to execute these good farm practices. Net farm income is gross operating margin (crop sales minus variable costs) minus the fixed costs, without accounting for family labour.

If good farming should pay-off, then farmers should be able to earn a sufficiently high net income from total farm activities, although income is not the only purpose of farmers. Like in any occupation, being a farmer is the expression of personal qualities and preferences, but also the expression of service delivery to others in a division of tasks within the society (egocentric versus altruistic). Being a farmer is a mix of self-realisation and service delivery.

> **Soil depreciation allowance**
>
> Faeth *et al*. (1991) did some research at farm household level. They subtracted not only depreciation costs for man made capital assets from the gross margin, but also a soil depreciation or appreciation allowance (SDA), which is an estimate of the present value of future income losses due to the impacts of crop production on soil quality. Soil erosion causes loss of plant-available water, and (thus) of soil crusting that restricts seedling emergence and root penetration, and in that way loss of plant nutrients. SDA measures the productivity of soil and not of technology. Therefore, the important issue is whether the technology is more productive on a better soil than on a degraded soil, to determine whether or not some of the effects of soil degradation have been masked by technology improvements. If the latter is the case, then there has been a real loss of income. Gross operating margin and government subsidies were added together and then the SDA was subtracted from this. The Net Economic Value (NEV) was calculated by subtracting from the net farm income an additional amount for the off-site environmental costs (sedimentation, effects on fisheries and recreational areas, downstream water users). Therefore, the NEV takes into account the costs that farmers' activities impose on others, but excludes transfer payments to farmers such as subsidies that are not income generated by agricultural production.

A farmer or farm family could prefer the combination of on- and off-farm activities, not only for economic reasons but also for social reasons. In the Netherlands, only 30%-35% of the total farm households earns their income purely from their own agricultural production.

This means that in 70%-65% of the farm households, part of the income comes from outside their own farm (Van Broekhuizen and Van der Ploeg, 1997).

Wiskerke (1997) concluded from his research in Zeeland, a Dutch province, that at full crop production farms only 55% of the total income comes from real crop production activities. De Vries (1995) studied the composition of total income of farm families in a Dutch region called "Land van Maas en Waal". Her figures show that total farm income from on-farm agricultural activities was 65% in 1987 and 63% in 1991. In 1987, 45% of the farm households in this region had a farm income, which came purely from on-farm agricultural activities. In 1991, this figure was reduced to 36%. Vereijken (1994) suggests the net farm surplus of sustainable farms should be > 0 (turnover minus all costs).

2   *Total farm family income*

To warrant a viable landscape's sustainable management, sufficient farm families should have a potential to remain farming in the rural areas. Total on and off-farm activities should be high enough to survive. If the total net farm income does not meet the needs of the family's expenses, additional income is needed to survive on that farm in the region. The existence of farm families in the rural areas is a prerequisite for the production of sustainable agro-landscapes. If good farming alone does not give enough cash income to survive, a farm family could increase its income through off-farm activities. Also other reasons, than financial or economic ones, could stimulate a farmer or farm family to fulfil on-farm as well as off-farm activities. As mentioned before, being a farmer with on and off-farm activities is the expression of personal qualities and preferences, but also the expression of service delivery to others in a division of tasks within the society (egocentric versus altruistic). The survival of farm families, and thereby their farming system, is a prerequisite for the survival of sustainable landscape management. Total farm family income is the sum of total net farm income and the income from off-farm activities.

The potential of a farm family to remain farming and thus to survive as landscape producing actor in the rural area, is not restricted to the on farm production of food and fibre alone. Multiple land use and off-farm activities improve the financial subsistence of a farm family. Total on- and off-farm income together should be of a certain level, compliant on the regional income per capita.

Examples of on-farm multi-activities that increase the added value and farm income are:

- Management and production of nature and landscape;
- Production of high quality products such as regional products, ecological and bio-dynamic products;
- Marketing and selling of products directly from farms;
- Agro-tourism and farm camping;
- Non-farm activities in former agricultural buildings such as storage facilities;
- Farms facilitating care-tasks for people who need care.

Possibilities to increase on-farm income are a low-cost strategy of farm activities and co-operation between farms at local and/or regional level. Recent studies conclude that different ways or styles for farming can all be economically viable and ecologically sustainable in the short term as well as in the long term (De Bruin, 1997; De Vries, 1995; Kerkhove, 1994; Van Broekhuizen and Van der Ploeg, 1997; Van den Ham *et al.*, 1998; Wiskerke, 1997).

Aspects, which affect farm income are: number of labourers, farm size, scale, level of self sufficiency, added value on products, level of mechanisation, level of cash flow, level of external inputs, and level of financial autonomy.

For instance, Kerkhove (1994) calculated for different farming styles the impact of combinations of the above mentioned variables and concluded that an increase in purchased external inputs together with a higher level of mechanisation has a negative impact on labour income.

Integrated farming pays-off

Van den Ham et al. (1998) show that farmers have different strategies to reach various purposes. They distinguish (1) integrated-farmers, including organic farmers, who focus on the integration of all aspects of farm management including landscape and nature production and (2) production-farmers who focus mainly on few or single crop or milk production. They concluded that integrated-farmers have more possibilities to include landscape and nature production into their farming systems than production-farmers have. They also concluded that the low cost strategy of the integrated-farmer gives higher net farm profits, return on labour, farm income and family income. These financial results are caused by their strategy and not explained by their additional income from landscape and nature production. This means that higher investments, specialisation, and especially scale-increase are not the only possibilities let alone the guarantees to improve farm income. One could say that income diversification instead of full income from food alone could be a criterion for professional farming. Especially conventional farmers, economists and agricultural policy makers think that scale enlargement will solve the financial and economic problems within the current agricultural sector. However, by definition, the consequence of scale enlargement is that many farmers have to disappear, together with their farms and their farm's landscape and infrastructure.

---

3   *Return on labour*

Generally, return to labour is measured as labour productivity, which is the total production per labour unit. Traditionally, wages for farm workers are treated as costs to the farm business, while returns to family labour are derived from profit. For farm management purposes this may be appropriated. However, the return to total labour (labour income) is a more useful measure in assessing the contribution that different farming systems can make to rural employment and incomes (Lampkin and Padel, 1994). It is calculated by adding the income of the farm family to expenditure on hired labour after any necessary adjustments for real and notional returns to land and capital (Lampkin and Padel, 1994). Bateman (1993) noticed that labour income represents the return to total labour, paid and unpaid, and provides an indicator of the contribution which different farming systems make to societal income and employment objectives. Favourable returns to labour on good farming systems, stimulates the continuation of such good farming practices and thus of sustainable landscape management. Besides that, it will improve the well being of living in the rural area and it will favour the rural employment. Here, return to labour can be measured as: (1) total farm family income divided by total amount of working hours, distinguished to on- and off farm labour hours, (2) total farm profit divided by total on farm labour units.

The effect of multi-activities on the return to labour depends on the available amount of labour at farm household level. If there is unused farm labour available, then the extra added value will increase farm income and labour productivity. If these multi-activities ask for extra (external) labour, then the extra added value may increase farm income and also local and/or regional employment. One should keep in mind that focussing on only an increase in crop and/or animal production often have already crossed the level where increase in inputs are not efficient anymore. This means that the increase in costs of inputs (labour, fertilisers, mechanisation, etc.) to produce extra units of milk or crops are higher than the increase in outputs. Therefore, it is useful for farmers to know their labour productivity distinguished to their various on- and off-farm activities.

4   Farm's market orientation

The level of market orientation says something about the possibilities to earn cash income. Farmers and communities should try to find a balance between market orientation and subsistence farming so that they are able to survive in the rural areas and so their sustainable farming practices. Farmers and co-operatives which are completely depend on anonymous markets for their outputs as well as for their inputs are confronted with low output prices and high input prices, which have led to high capital on-farm investments. This may result in farmers eating into their own capital. Examples of such farmers are often found at capital intensive farms such as intensive chicken farms, pig farms, and intensive crop production farms. To avoid or stop this process, farmers should create added values on products from their farm based on farm and off farm activities. Purely subsistence farming, which means without any cash crops and without any off-farm activities, does not fit in the current socio-economic systems. The economic and social structures of the rural areas require a certain level of cash income from farm families.

The level of market orientation and the possibilities to earn cash income without eating into the farm capital depends on the style or type of farming and/or the infrastructure of the region. If good farming should pay off, then capital involvement and decision making have to be considered: actual farmers as owners or as hired staff for land owners versus decision making. Farmers with a passive market orientation produce mostly bulk products, which are sold in the market at low prices. The short-term advantage of such farming activities, for especially the agribusiness, is low food prices. Farmers with an active market orientation are willing and able to adapt their activities towards consumers and society's preferences (empathic coherence). Those farmers create an added value to their farm based on on- and off-farm activities for which consumers are willing to pay for. An active market orientation also means to be active on various markets and not only one market. It can be said that conventional agriculture reacts slowly at changing preferences of consumers and society and that conventional farmers focus mainly on one market and are almost completely depend on input and output markets.

The level of market orientation can be measured as the ratio of the total production costs and the value of reused on-farm products. A high ratio means that the farm is highly market oriented. Another aspect of the market orientation is the possibility of a farmer to create an added value towards their farm-related products, which are traded in the market. In general, ecological and multi-functional farmers create higher added values to their products than conventional farmers (Kerkhove, 1994; Van Broekhuizen and Van der Ploeg, 1997; Van den Ham et al., 1998; Wiskerke, 1997). These farmers are more flexible, which means here that they are able to react on consumers' and society's preferences (emphatic coherence). For this aspect of the market orientation, the share of added value in total net production value can be calculated.

5 Financial autonomy

Next to farmers' dependency on input- and output-markets, are farmers getting more and more dependent on financial networks to support their farm. A farming system based on high investments in land, capital and external inputs requires large amounts of money, which farmers do not have. The 'solution' is given through bank loans. The farmers' financial autonomy is decided by the level of invested capital and thus by the dependency of external money suppliers. As productivity increases, so does capital investment. The level of financial autonomy can be measured as total net farm income divided by the value of total invested capital for agricultural as well as non-agricultural related farm activities (Tellarini et al., 1996).

Financial autonomy makes farmers and farm co-operatives less dependent on the financial networks such as banks. Dutch farmers have been stimulated to produce high quantities of products at standard qualities, in order to have cheap food for everybody. As mentioned several times before, this policy also led to low output prices, high input prices, environmental degradation, and high capital investments in order to make the production process more efficient (higher outputs with less inputs). At the same time, farmers are confronted with a policy, which force them to make high capital investments to reduce the environmental negative effects from their farm practices. In order to make the investment pay, many farmers have kept investing in new technology an increasing the size of their farms. This practice tends to be self-reinforcing (Greenpeace, 1992). Taking over a farm from one generation to another one goes together with the creation of loans to finance the take over of the farm. This means that banks are making good business just by waiting that the new generation is ready to take over the farm and comes to the bank to ask for finances. In all cases farmers are not able themselves to pay these necessary capital investments or the take-over of a farm. They are getting more and more dependent on banks giving loans on security, which means that many farms belong to banks. The value of farms can be enormous, depending on the location, the farming activities, and the regional policy. Banks have generally benefited from the Common Agricultural Policy (CAP).

To a great extent banks provide farm capital. The tread-mill effects of this policy and farming systems reduces farmers' financial autonomy and is mostly related to ecologically and economically unsustainable farm practices and landscape management.

Financial autonomy and financial burden in agriculture

Between 1960 and 1985 total capital input in Belgian agriculture increased by ± 550% (in current prices). The net burden on farms in Belgium, doubled between 1973-1975 and 1978-1980, accounting for 18%-36% of net farm income (Greenpeace, 1992). For many farmers in Europe, farming does not pay its way any more (Greenpeace, 1992; Kerkhove, 1994; Van Broekhuizen and Van der Ploeg, 1997; Wiskerke, 1997).

3.2   Main criterion: Greening the economy

MAIN TARGET: Economic calculations at farm level and at regional level should include environmental costs and benefits. As mentioned by Daly *et al.* (1990) and Tellarini *et al.* (1996), identification and (re)allocation of costs is very important in this greening process. Examples of greening incomes and accounts at farm level are the aforementioned studies of Faeth *et al.* (1991), Defrancesco and Merlo (1997), and Tellarini *et al.* (1996). Greening the economy in such a way that external costs and benefits of farm activities are finally included in farm produce prices can not be done by the farmer alone. (Regional) policy and support from the local community will be necessary to stimulate the greening of the economy. There are various possibilities to internalise the ecological costs and benefits of land use activities. See for instance Van der Lubbe (1996), who gave an overview of the various methods to value and internalise ecological costs and benefits. Although our society is nowadays perceived as highly economised, all these external benefits from multiple land use are not yet included into the neo-classical, mercantilist, (semi-) capitalist economy prevalent in "western" society. As this approach to economy is oriented toward externalisation of production costs for the private entrepreneur, internalisation of environmental and social costs does not come easy. It is opposed to the "economic" way of thinking and the compliant attitude. So from this point of view, economics is not yet sufficiently radical and general to warrant the above values. Respect for ecological potentials and cultural achievements, in view of technological possibilities, is not sufficient any more to warrant their survival (Hueting *et al*, 1992, Daly *et al.*, 1990; Daly and Townsend, 1993).

One approach to the reduction of pollution from agricultural sources and to the internalisation of environmental externalities within the costs of production is to introduce taxes and charges on potentially polluting agricultural practices or inputs. Such taxes and charges are often referred to as 'economic instruments' a term, which may also include incentive payments and subsidies that already are a major component of agricultural policy (Baldock and Beaufoy, 1993).

Much of what is valued in the rural environment is a legacy of a historical pattern of an agriculture based social economy. Much of the (re-) newly appreciated bio-diversity has been developed as part of the multi-functionally differentiated agricultural land use that shaped the landscape. However, the recent changes that have taken place in the organisation and technology of agricultural production, separating the functions and allocating them for example in profitable and marginal regions, have undermined the apparently complementary relationship between agriculture and the rural environment (empathic coherence). As long as people are not aware or do not feel responsible for such external costs as mentioned, they will not be internalised. This means that they will not be paid for by the actual consumers (beneficiaries), but will be rolled down viz. removed either to future societies or societies elsewhere. Thus one can say that current economy, at macro as well as at micro level, expresses society's alienation from the external costs and benefits coming from rural, agro-sylvi-pastural land use and landscape in general (Hueting *et al*, 1992).

In recent years, many NGO's and later UN and EU delegates have started looking for a more fysiocratic, environment and public interest focused approach of the economy, in which external costs and benefits are internalised. That is: incorporated in the production costs and thus in decision making of producers and consumers. This then is instrumental in adjusting society's behaviour towards a sustainable mix of production and consumption (Bruntland, 1992; Greenpeace, 1992; Panayotou, 1995; Von Weizsäcker *et al.*, 1997).

Parameters for greening the economy:

1   Technical autonomy
2   Dependence on non-renewable inputs
3   Share of re-used on-farm production value in total costs
4   Share of non-renewable inputs in total costs
5   The costs-benefits ratio of investments in landscape, environment and nature

1   *Technical autonomy*

In general, conventional farming systems reduced their technical autonomy through their dependency on external inputs and high capital investments. The basis of this process has been the agricultural policy and technology development focusing on production efficiency without considering their external effects. This led to such situations where farms became more like industrial based processing units of raw materials than land based primary production units. Farmers got "transferred", through the ongoing specialisation process, from self-employed entrepreneurs towards tenant farmers who are completely dependent on contracts with the industry. Such farmers function just as a small part of the food production chain. Well-known examples of such industrial based processing units are intensive breeding and meat production. Those farms do not have a positive impact on the landscape farms (Volker, 1997). Most of those farms produce awful smells and are associated with unfriendly animal live. They may also have a negative impact on the region. For example, the pest disease on pig farms in the south of the Netherlands led to negative associations of regional and intensive farm practices. Technical autonomy is measured as the share of the farm activities in the large-scale production process or production chain. Another way to measure this parameter is the ratio between own labour and contracted labour. This parameter refers to the amount of on farm activities, which is executed by contracted or custom labour. Finally, the ration between investments in machines and labour can be used as parameter for the technical autonomy.

2   Dependency on non-renewable inputs

In order to improve farm management towards sustainable land use and landscape management the use of external, non-renewable inputs should be minimised. Non-renewable inputs have to be brought from outside the farm and are by definition not sustainable in ecological as well as economic terms. This means that they are not part of sustainable farm management. The dependency on non-renewable inputs can be derived from farm management data and is calculated as the ratio between non-renewable inputs and total inputs.

3   Share of re-used on-farm production value in total costs

On farm produced products can be re-used at the farm. For instance, as feed for animals, food for the farm family inputs for food processing, construction materials, fuel wood, soil fertility, etc. As with renewable inputs, re-use of on farm produced products stimulates a greening economy at farm level as well as at regional level. An increase of re-used farm products reduces the production of external effects and of waste at farm level, but also at regional level. This parameter is measured as the ratio between the total value of re-used farm products and the total production costs.

4   Share of non-renewable inputs in total costs

This parameter is closely related to the second and third parameter under this target. Here, the dependency on non-renewable inputs is expressed in monetary terms and is calculated as the value of non-renewable inputs divided by the total production costs.

5   The cost-benefit ratio of investments in landscape, environment and nature

Investments in landscape, environment and nature will have a positive impact on landscape. If the investments are also efficient for a farmer or farm co-operative depends on the receipts. This parameter is expressed in monetary terms and evaluates if the benefits, mostly received as subsidies, are higher than the investment costs. A cost-benefit ratio smaller than one makes it worthwhile for a farmer to invest in landscape, environment and nature.

3.3   Main criterion: Regional autonomy

MAIN TARGET: Regarding the majority of the region's production and consumption of bulk and staple food, fibre and energy, the rural region's subsistence based in agriculture, fishery, and/or forestry will be necessary for regional autonomy. There should be certain potentials for agriculture, forestry and fishery within a region, as basic socio-economic and geographic carriers for sustainable landscape development. Obviously, wherever possible, the region's food and fibre surpluses can and should be used to serve neighbouring urban areas. Regional (rural) development policies can play an important role in favour of this regional autonomy. A further process of specialisation and the excessive urbanisation (to macro and mega-poles) has led to the alienation between the rural area's farmers, demanding fair prices, and city's consumers demanding cheap food. Fixation on the fancy of standard quality of ready made uniform end-products lead away from the awareness of the region and season bound agricultural production process, and thus also from the regional origin of quality food.

Parameters for regional autonomy:

1	Transport
2	Resource efficiency and regional labour possibilities
3	Swaps from single commodity support to management system's support
4	Translation of the parameters under main criterion 3.1 and 3.2 to regional level
5	Market access for regional speciality produce

1   *Transport*

A rural area with a high level of transport activities from outside the region means that there is hardly any regional autonomy. As has already been mentioned before, enormous amounts of inputs, outputs, raw materials and waste are transported all over the world at the costs of enormous amounts of energy. A reduction in transport will reduce energy losses and will stimulate regional autonomy. The level of transport to and from the region for farm activities can be measured by counting the transport activities to and from the farms. We, "the raw materials society", have to keep in mind that, to a great extent, surplus production in Western Europe is possible only by using raw materials from elsewhere. Intensive livestock rearing in particular depends on large imports of "cheap" protein-rich feed stuff mainly from Thailand, Argentina and Brazil. Some 38% of the world's grain is now fed to livestock. Of this, pork production uses more grain world-wide than any other meat industry and the EC alone produces 20% of the world's pork (Greenpeace, 1992). Together pigs and poultry account for two thirds of feed-grain consumption. Apart from the global social and environmental impacts of intensive animal production in Europe, the import of protein-rich animal feed is one of the major contributing factors to the disruption of the nutrient cycle on European farms. These animal feeds constitute a major import of nutrients and energy to the agricultural system in Europe while soils in developing countries are being depleted of their nutrients. Imported animal feed is one of the main contributors to the millions of kilograms of nitrogen that accumulate in the Netherlands every year. Ironically, attempts to solve the nutrient problem in the Netherlands and other countries have included processing (drying) the purposely wetted animal waste to allow transport action back to the countries that exported the animal feed. This attempt to 'close the nutrient cycle' at a global level demands huge amounts of energy from non-renewable sources, on top of the energy already used to transport the feed. It is far from an ecological or sustainable solution (Greenpeace, 1992).

2   Resource efficiency and regional labour possibilities

One of the reasons for rural development is the sociologically devastating urbanisation. National and regional policy, considering resource efficiency and optimising labour possibilities are necessary to support regional autonomy in agriculture, fishery and forestry. We have to keep in mind that the surplus production in Western Europe is possible only by using resources from elsewhere, such as land in Third world countries for the production of animal feed (grain and tapioca). Efficiency figures of resources will change considerably if the claims on resources from elsewhere are also taken into account (Bakker, 1985; Greenpeace, 1992; Marino et al., 1997).

Regional autonomy requires optimal use of local resources, including labour. This does not mean a further regional or local specialisation in searching for resource efficiency. On the contrary, optimal regional resource use is not determined by the economic meaning of efficiency only. Also ecological and social criteria play an important role in searching for an optimal use of regional resources. In order to reach and to stimulate a certain level of regional autonomy new jobs have to be created to keep young people in the rural area. For instance, ecological and bio-dynamic farming will increase labour requirements in the rural areas. Also non-agricultural on-farm activities may have a positive impact on job creation in other sectors. Every region has its own characteristics and specialities. Although such regional characteristics and specialities will improve resource efficiency, one should keep in mind that the region should avoid getting completely dependent on these characteristics and specialities.

3   Swaps from single commodity support to management system's support

Next to total on- and off-farm income, farmers and farm families are dealing with subsidies and taxes derived from agricultural policy at regional, national or European level. A swap from single commodity supports to management systems' supports by the regional policy will stimulate farming practices which are in harmony with the environment and ecology of the region (characteristics or "typical" products of the region) (empathic coherence). As mentioned several times before, the single commodity promoted agricultural systems resulted in low output prices, environmental costs, monotonous landscape and a reduction of employment in the rural areas. More and more countries believe that the protection and enhancement of the natural and cultural heritage of rural areas is a major objective of rural development policy (OECD, 1994). Also the creation of employment for rural development is seen as prerequisite for the survival of rural areas (OECD, 1995).

Baldock and Beaufoy (1993) concluded that no one policy instrument of the Common Agricultural Policy (CAP) of the European Community (EC) is adequate to provide improved protection for 'high nature value' agriculture on a European scale or to ensure that farm practices, appropriate for nature conservation, are retained or adopted. This fully complies with Witte *et al.* (1993), stating that there is no way to warrant sustainable agriculture by controlling the price of any one single major products, like for example rice. Of the policies reviewed by Baldock and Beaufoy, special attention was paid to the role of *cross-compliance* and positive environmental incentive payments. They mentioned that these measures can have a central role in providing a foundation for strengthening the viability of 'high nature value' agriculture in the medium term and supporting those farm practices of particular value for nature conservation. Well-adjusted support from national and regional government levels is needed to make sure that local initiatives to improve landscape management are encouraged (Michelsen, 1997).

Rønningen (1996) presented some main experiences with measures directed towards the rural landscape in Western Europe in recent years. Hectare payments and landscape management agreements are the two major measures financed trough agricultural budgets. It has been concluded that hectare payments may lead to reduced intensity in farming methods, however, without specified prescriptions, their importance in terms of environmental improvement are limited. Landscape management agreements for certain defined areas may be more efficient for reaching defined environmental aims. Another instrument, land consolidation, is directed towards the rationalisation of production and the 'dynamic' development of land and will often be in conflict with traditional ideas of landscape conservation. In Germany and Denmark, this instrument has been used to re-create bio-topes and landscapes - re-naturalisation. Registrations of valuable landscapes were carried out to serve as a basis for selecting areas for landscape management measures (Rønningen, 1996). If landscape has to be maintained or developed by farmers or farmer co-operatives then the question for society and policy makers is how this can be done by means of well-targeted farmer income support.

As mentioned several times before, national and/or regional policies have an important impact on farm management and thus on land use and landscape management practices of farmers and farm co-operatives. Single commodity support stimulates mono-cultures and the standardisation of farm practices. Management system's support makes it possible for national and regional policy makers to influence farm management. It reduces the exploitation of mono-cultures, the standardisation in the agricultural production and the related monotone landscapes. If well done, it will stimulate sustainable farm practices and multiple on-farm activities. The existence of management system's support or single commodity support can be derived from farm management data.

4   Translation of parameters under main criterion 3.1 (good farming should pay-off) and 3.2 (greening the economy) to regional level

The parameters mentioned under main criterion 3.1 (good farming should pay-off) and criterion 3.2 (greening the economy) of the Table of the Checklist (Table 3.1) should be further translated to a regional level. Here, measurement of farm income and related parameters should be interpreted at a regional context. The same can be said about the parameters of the greening process of farm economics. Those parameters should be interpreted and evaluated at a regional level, so that relative values of parameters can be compared within a region and between regions.

5   Market access for regional speciality produce

One of the aspects to support regional autonomy regarding bulk and staple food, fibre and fuel is to improve the market access for regional speciality produce for example wines, beers, cider, herbs, cheeses, meats, bakery products, sweets, etc. In wines or vinology, the links between site and quality is best known. Regional specialities give a good name to a certain region. If farmers are able to create an added value on the production of regional specialities, then they have a good reason to remain farming especially in that area. The added value of regional speciality products includes next to quality also cultural and psychological elements relating to the specific area. Also in the Netherlands, more and more farmers and farmer co-operatives or farmer organisations try to develop a regional speciality in order to distinguish their region's characteristic quality from others and to support regional autonomy. Farmers are looking for alternatives for the conventional agricultural system based on standard products, mono-cultures, low prices, and high investments. These bottom up initiatives are divers, but the common element is that farmers together are looking for new activities, which give an added value to their farm activities. Basically, such new activities are related to the positive values or characteristics of the region. Results from a recent inquiry show that farmers are not enthusiast anymore to follow the style of farming which have been promoted for years by the agricultural policy and which goes together with high investments, homogenous products and farmers anonymity (Van der Ploeg, 1997). A more social or psychological aspect of regional specialities, which can be used for a region's survival, is the re-appreciation of regional foods and fibre, see sub-criterion 1.1.1 (fertile and resilient soil) of Table 3.1. Also the regional landscape features, which invite all kinds of active and passive recreation such as sports, leisure, and (on-farm) landscape maintenance, can be used as an instrument for regions' survival (Giorgis, 1995). The production of regional specialities can be stimulated by regional and/or national policy. The existence of market access and the level of market access of regional specialities can be derived from farm management data. The region itself will also be able to supply information about its specialities.

4   SOCIOLOGY (Column 4)

Here, sociology refers to the social structures, which allow people to get access, to participate in and to take responsibility for decision making on landscape and land-use planning. It also refers to the implementation of landscape and land-use planning after the decision making process and to the social reproduction of viable and socially recognised life in the rural area, which is a prerequisite for the ongoing management of landscape (Volker, 1997). A leading idea is to re-establish participative connections, on a basis of equity, between farmers, other landscape managers, and all the other social partners involved in landscape planning, management and consumption. This means that all landscape stakeholders should be involved in the participative development of sustainable landscapes and thus sustainable land use management.

Four main criteria have been distinguished with respect to participative and social structures:

4.1	Well-being in the area
4.2	Permanent education of farmers
4.3	Access to participation
4.4	Accessibility of the landscape

4.1   Main criterion: Well-being in the area

MAIN TARGET: This target focuses on the conditions allowing for ongoing, acceptable life in the rural area and the rural landscape, which is a pre-requisite for social reproduction of sustainable landscape management. Although farmers themselves are not able to change easily the level of well-being in their region, it plays an important role in the possible re-production of sustainable landscape management. It is not only the farmer him/herself who is looking for well-being, but also the farm family. For instance, farmers' children, like all other children in the region, will leave the area if there are no welfare service available. The same can be said about retired farmers. Improvement of the well being of an area counteracts the rural degradation by increasing the social viability of the agro-landscapes. It acts as an incentive for rural development, which again supports de-urbanisation (Volker, 1997; Beissman, 1997; Bosshard, 1997; Colquehoun, 1997). Viable and appreciable life in the rural areas is seen as an important counterbalance to the excessive urbanisation going along with the devastating exhaustion and desertification of rural landscapes and the resources. The question is, which social conditions are necessary to come to viable and appreciable life in the rural areas linked up with sustainable agro-landscapes? Options for farmer's succession and a certain level of farm income and welfare services in the area are necessary social conditions.
Note: Here again, the land property and land-management structures from farm to regional level are crucial.

Well-being in the rural area: job creation (OECD, 1995)

Rural areas in many European countries get empty since young people leave these regions hoping to find a living in cities, because the rural areas do not offer them good possibilities for e.g. jobs. The OECD (1995) studied the problems of job creation in rural areas in an effort to contribute to the wide range of problems caused by over-urbanisation. Three broad categories of problems were distinguished:
(1) Information problems, such as concentration of information in metropolitan areas, distance from the financial networks and decision centres, and difficulty in cultivating contacts for business purposes;
(2) Human resource problems, such as demographic decline and youth exodus, limited human resources, with little possibility for getting access to training and absence of industrial and third party work experiences in areas dependent on agriculture;
(3) Small scale problems, such as relative small local market, scarce economic and social services, and inadequate infrastructure.
See also Marino *et al.* (1997).

Parameters for the well-being in the area:

1   Options for farmers' succession
2   Financial income
3   Welfare services in the region

1   *Options for farmers' succession*

Options for farmers' and farm management successions, in order to let 'good' farmers remain in or move to the rural areas in need of good agro-landscape management, have to become feasible and realistic. The possibilities for farmers to retire with an acceptable pension from savings or selling the farm to successors play also an important role here. The sales option requires good income perspectives for the successor. See column 3 (economy) of Table 3.1 (Table of the Checklist). It should be possible for new generations to become a 'good' farmer without an unbearable financial burden. However, retiring farmers, without (extra) pensions, need enough money for their future without any incomes. If they can not get a pension from selling their farm, they must stay on the farm or find other solutions.

Next to the financial aspects of taking over a farm, there are also the social aspects, such as the difference in opinion about farm management between the old and new generation, the (landowner's) family relationship, the (landowner's) family tradition, etc. The next main criterion (permanent farmer's education) pays attention to the knowledge of farmers about sustainable farm management in order to succeed sustainable farm management or to start with it if so needed (conversion from less-good to better farming systems).

2   Financial income

As already mentioned in column 3 (economy) of Table 3.1 sufficient financial income is one of conditions to remain farming and thus to make social reproduction of sustainable landscape management possible. For further details of this parameter, see main criterion 3.1 (good farming should pay-off).

3   Welfare services in the region

Other conditions, which determine the level of well-being in an area are the existence of welfare services such as shops, schools, clubs, churches, pubs, transport, etc. These welfare services together with job possibilities form the institutional framework or infrastructure for the well-being in the area. However, the farmers themselves have less influence on this parameter than on the other parameters in this study, although these services have a main impact on the well-being of farmers. The well-being of farmers in the area depends to a great extent on the context in which farmers and the rural community as a whole, live and operate and with which they have to deal without any other choice. Some of the welfare services are a prerequisite for people to make life in the countryside possible and comfortable. The existence of welfare services may have influence on the off farm activities of the farmer.

4.2   Main criterion: Permanent education of farmers

MAIN TARGET: Next to the well being in the area, which can be compared with Maslov's first stage of physical survival, farmers should be aware of sustainability issues in land use viz. agro-eco system management. If farmers are not aware and do not have the appropriate education, it will be difficult for them to execute sustainable land use and landscape management. Therefore, this criterion refers to the development of farmers' knowledge about sustainability issues. Farmers have several possibilities to improve and to develop their knowledge about issues such as ecological and bio-dynamic agriculture, landscape management, etc. Farmers' knowledge about sustainability issues depends on their level of education: secondary school and professional education. After they became farmers, there are many possibilities for farmers to improve and develop their knowledge on sustainable land use and landscape management, for instance participation in sustainable agriculture and landscape relevant study circles, training, and courses. Development of farmers' knowledge through education can be indicated as the stage of self-realisation in Maslov's theory.

Parameters for permanent education of farmers:

1   Farmer's level of education
2   Farmer's participation in sustainable agriculture and landscape relevant study circles, training and courses

1   *Farmer's level of education*

This parameter is straightforward and measures the level of education that the farmer has passed through. We presume that the higher level of education, especially to subjects of sustainable agriculture and landscape management, the higher farmers' awareness about and initiatives on sustainable landscape management (Estupindin *et al.*, 1997).

2   *Farmer's participation in sustainable agriculture and landscape relevant study circles, training and courses*

There are many possibilities for farmers to develop their knowledge about sustainability issues after they become farmers. Examples are study circles with colleagues, special training courses about landscape and landscape management, visits or excursions to colleagues, etc. This parameter will also indicate farmers' knowledge and awareness of sustainable landscape management and can be measured through farmer's participation in sustainable agriculture and landscape relevant study circles, training, and courses.

4.3   Main criterion: Access to participation

MAIN TARGET: The aforementioned landscape degradation is seen as a result of alienation between the various stakeholders as such and between the stakeholders and the values of nature and rural landscape. Increased awareness of the mutual interests and interdependence of farmers, the rural community, urban people, and government will give incentives to a multiple range of possibilities to co-operate (empathic coherence). There seems to be a lack of information, knowledge, understanding, and awareness among farmers as well as policy makers, extension services, researchers, consumers, and laymen about sustainable landscape management and the role each other plays or can play in this common interest. Access of farmers to participate outside their own farm and the involvement of outsiders to farm activities will increase the social acceptance of farmers and of their farm activities. Farmers' involvement in landscape development requires support from the local community (including consumers). On the other side, the fact that local community or local stakeholders, from consumers to planners pay farmers' involvement in the landscape requires farmers' participation in local communities' decision making processes (Beissmann, 1997; Bosshard, 1997; Volker, 1997). The common interest of farmers and local community can merge in win-win solutions:

farmers are paid for landscape production, which is appreciated by the local community, including consumers (empathic coherence). Here is a possibility of the before mentioned cross-compliance. Communication between farmers and the community will reduce the existing alienation gap between farmers and the community. Farmers and the local community together are responsible for the landscape of their region. This means that a farmer can do with his land what he wants till certain limits, which are set by the community. Bennett *et al.* (1990) mentioned that adjacent communities with a clear understanding of control over their land or forest (or aquatic domain), be it private or communal can negotiate conflicts arising from externalities. The involvement of local stakeholders in planning, decision making, and landscape management of their area, gives them access to or influence on decision making and landscape management of farmers or co-operatives in their area and vice versa. Access to participate in local community and in farm activities will increase responsibility among farmers as well as among local community for sustainable landscape management. There are various ways how these two way participation and involvement can take place, e.g. through the membership of farmers in regional councils, farmers' organisations, co-operation with NGOs and consumer groups, professional and lay excursions to the farm, etc.

This main criterion is subdivided into the following sub-criteria:

4.3.1   Farmers' involvement in activities outside their farms
4.3.2   Outsiders' involvement in farm activities

4.3.1   Sub-criterion: Farmers' involvement in activities outside their farms

TARGET: Participation of farmers outside their farm can take place at various levels, from colleagues to governmental level, and will increase farmers' awareness and willingness on sustainable landscape management. Off course farmers' willingness to co-operate with other landscape stakeholders is crucial and should be encouraged to warrant the viable and sustainable development of landscapes (Beismann, 1997; Bosshard, 1997; Colguhoun, 1997). The point of this target is: "Where do farmers have access to participate in activities outside their farms, which will increase their awareness and willingness on sustainable landscape management?"

Parameters for farmers' involvement in activities outside their farms:

1   Membership to farmer organisations and farmer groups
2   Working in the region
3   Involvement in organising outlets
4   Co-operation with NGOs
5   Membership of regional councils
6   Access to professional expertise and support programme
7   Access to participate in dissemination programs

1 Membership to farmer organisations and farmer groups

Farmer organisations and farmer groups are groups of farmers, which co-operate on various subjects, for instance co-operation on environmental or landscape aspects. Such organisations or groups, existing of colleagues are an important information source for a farmer. The opinions of colleague-farmers have considerable influence on the awareness and activities of a farmer. A farmer may improve his knowledge on sustainable land use and landscape management through his membership to a farmer organisation.

2 Working in the region

Off-farm income is not only relevant for main criterion 3.1 (good farming should pay-off). Here, off-farm income is related to labour diversification and social integration of farmers. Off-farm activities increase farmers' knowledge and awareness about what is going on within the community and the region. Farmers get to know what is expected from them as farmers with respect to sustainable landscape production. This parameter can be measured by the ratio between the time spend on on- and off-farm activities or between on- and off-farm income.

3 Involvement in organising outlets

Farmers participate in the development of a market for sustainable agriculture and landscape through their involvement in organising outlets, in which they try to co-operate for the selling of their products. There are various ways in which farmers can organise outlets of their farm products. Examples are: selling products on local (farmer) markets, agreements or contracts with restaurants to deliver farm products, co-opeation with consumer and participation groups on food subscriptions for farm sales, selling directly from the farm and community supported/shared agriculture. For CSA's see Groh and MacFadden, 1990; Salm, 1997). Farmers and consumers do co-operate strongly together in CSA's. See also sub-criterion 4.3.2 (outsiders' involvement in farm activities). The involvement of farmers in organising outlets is not only to sell their products but also to inform people about the way they produce food and fibre. Also information about regional specialities can be spread through consumer and participation groups. Re-appreciation of food and fibres, but also of regional landscape features that invite all kinds of active and passive recreation, are instrumental to the regions survival (Giorgis, 1995). The involvement of farmers in the various types of outlets can play an important role here. Farmers' participation in all kinds of outlets will improve their knowledge about consumers and what they expect from a "good" farmer to produce, viz: "good" farm products including landscape. Participation of farmers in the various types of outlets will reduce the alienation gap between farmers and the community. Working together will improve understanding between the community and farmers in such a way that farmers know what the community wants and the community knows what a farmer needs and can do in relation to sustainable landscape management (empathic coherence).

4   Co-operation with NGOs

Farmers may co-operate with different types of NGOs in order to improve their knowledge and awareness about sustainable landscape management. Farmers can participate in NGOs, which are active, for example, in nature conservation, wildlife, birds, flowers, but also more generally in environment and countryside programmes or eco-tourism.

5   Membership of regional councils

Farmers membership of regional councils will give them access to decision making in regional planning about the development of their region, including the landscape. Farmers' membership of regional councils will improve the understanding on the opinion, wishes, and possibilities of sustainable landscape management between local community and farmers. All local stakeholders together, represented by a good regional council, will feel responsible for their region, including the landscape. As with the other parameters, co-operation between farmers and other community stakeholders will improve their knowledge and awareness about sustainable landscape management.

6   Access to professional expertise and support programs

Farmers' access to professional expertise and support programs will supply them with the necessary information about the adaptation of the conversion from conventional agriculture to ecological or bio-dynamic agriculture. It will also increase their willingness for sustainable land use and landscape management practices. Off course, this parameter depends on the available professional expertise and support programs. The availability of expertise and support programs is for instances the responsibility of the government, NGOs, extension organisations, and research and development organisations. Moreover, "good" farmers can also play a role in the availability of expertise. This aspect is included in the next parameter.

7   Access to participate in dissemination programs

The participation of 'good' farmers with relevant information and experiences in dissemination programmes can be important and effective for adapting the conversion towards sustainable agriculture and sustainable landscape management. Such information can be interesting for colleagues, extension services, and research and development. This parameter can be assessed by considering farmer's participation in dissemination programs and by the accessibility of the farm for excursions, experiments, etc.

4.3.2 Sub-criterion: Outsiders' involvement in farm activities

TARGET: Participation of outsiders in farm activities can take place at various levels, from colleagues to governmental level, and will increase farmers' awareness and willingness on sustainable landscape management.

Parameters for outsiders' involvement in farm activities:

1   Access to participate in landscape management
2   Professional and layman excursions to the farm
3   Community supported/shared agriculture (CSA)
4   Financial commitment to landscape programmes
5   Access, given to farmers, to buy/rent and manage landscape in a sustainable, ecologically and socially sound way

1   *Access to participate in landscape management*

Here, the involvement of the local community in farm activities is considered. The question is if farmers do accept volunteers to work on their farms. Some farmers see these volunteers as interference in their farm activities. However, volunteers are motivated people who are willing to help a farmer to produce and to maintain sustainable land use and landscape. They may save a lot of work for farmers, at a low cost. The volunteers, on the other side, appreciate and enjoy this type of work and may learn a lot from the farmer. Access of local community to participate as volunteers in landscape management can help a farmer to develop and maintain sustainable landscape. The parameter can be considered by assessing the accessibility of the community to farm activities or to landscape management activities.

2   *Professional and laymen excursions to the farm*

Professional and lay excursions to farms make it possible for outsiders to get informed and aware about farm activities. Society is invited by the farmer to have a look at his farm. As mentioned in one of the parameters of sub-criterion 4.3.1, farms can be visited by colleagues, extension workers and research and development institutions, in order to get informed about the farm activities and sustainable landscape management. The information exchange between professionals and farmers can be organised through excursions to and/or experiments on the farm. Excursions for laymen play an important role in reducing the alienation gap between producers (farmers) and consumers (laymen).

3   *Community Supported/Shared Agriculture (CSA)*

Community supported/shared agriculture (CSA) directly links farmers and community (Groh and MacFadden, 1990; Salm, 1997). The local community participates in farm activities and farmers participate in community activities. It creates a strong relationship between the local community and farmers. CSA is organised in such a way that farmers are producing directly for the local community. CSA could be a good basis for taking co-operative responsibility for sustainable landscape management.

4   *Financial commitment to landscape programs*

Farmers' involvement in landscape development requires (local) community support. So far, management agreements are the most popular form of financial commitments to landscape programs. These management agreements focus on farm practices, which will have an impact on the landscape. It seems that farmers are willing to change their farm management if they receive financial compensation or a reimbursement. If the community is willing to pay farmers for the production of landscape, then farmers are willing to change their farm practices and landscape management.

5   *Access, given to farmers, to buy/rent and manage landscape in a sustainable, ecologically, and socially sound way*

The community or society should make it possible for 'good' farmers to buy or rent land and to manage the landscape in a sustainable, ecologically and socially sound way. For example in Switzerland can milk quota only be sold to bio-dynamic farmers. Another example has taken place in The Netherlands where land in the Flevopolder has been destined for bio-dynamic farming.

4.4 Main criterion: Accessibility of the landscape

MAIN TARGET Public accessibility of a landscape, site or farm is necessary to let the community experience and appreciate or 'consume' the landscape as such. If farmers do participate in the decision making process and responsibility of the local community, then this local community wants to 'consume' for what they are paying for. Besides that, if the local community is satisfied with and appreciate the landscape, site or farm, then they are willing to pay farmers for their activities in land use and landscape management. Accessibility of the landscape facilitates commitment of consumers to farming practices and landscape management. Farmers want and need appreciation for their products. But farmers are not always enthusiastic about the accessibility, because they feel disturbed in their privacy when people cross their farmyard and their land. Sometimes accessibility of the land leads to damage to fields and animals. Alienation and a lack of communication and information between farmers and the community could be the result of farmers' resistance towards accessibility of land and landscape or the cause for such damage. Such a situation creates a vicious circle. In former days, public access to "church-paths" and "rights of way" were general accepted by farmers and the local community. Farmers that become initiative in participating trough excursions of consumers, NGO, etc. to their farms may reduce the threshold of resistance towards accessibility and will facilitate commitment of consumers to their farm practices and land use and landscape management.

Parameters for the accessibility of the landscape:

1   Excursions to the farm
2   Right of ways
3   Tracking roads

1   *Excursions to the farm*

Farmers may invite landscape consumers to their farms so that consumers are able to enjoy and appreciate the landscape and farmers are able to control the visits to their farm and sites.

2   *Right of ways*

The existence of the rights of ways is the official public accessibility to farms, sites and landscapes.

3   *Tracking roads*

Tracking roads, such as walking paths, can be used by countryside walkers and for eco-tourism, which enjoy and appreciate the landscape. This means no forbidden entrances and good passages, such as swing-gates and easy step-overs for fences.

3.2.3 CRITERIA FOR THE HUMANITY REALM: PSYCHOLOGY AND PHYSIOGNOMY/ CULTURAL GEOGRAPHY

J.D. van Mansvelt and J. Kuiper

Introduction

After the criteria for qualities of the bio-sphere and the social sphere have been treated in the previous subsections, now the turn is on the criteria for the qualities of the cultural sphere of history, architecture, anthropology, aesthetics and ethics, as reflected in and addressed by the landscape. As pointed out in the introduction of this chapter, and particularly in chapter 2 about the methodology, this sphere is, in line with the Maslow approach, perceived as the last in priority when physical survival is at stake. However, at the same time it is the realm reflecting the ultimate goal of humane development, with all previous disciplines and aspects mentioned in the Table of the Checklist (Table 3.1) acting as facilitators of these ones. After physical and social survival being sufficiently warranted, it is ultimately humanity's ethical survival that is at stake in any truly humane development (UNCED, 1992).

Here, the landscape, expressing its underlying history of agro-sylvi-pastural-land-use, from the consumer's point of view, is approached from its least material side, in its most non-excludable and non-depleting product form. "Landscape is seen as the total and perceivable appearances on the surface of the earth, which result from the interaction between man and nature" (LNV, 1992). At large, its colours, smells, structures and its picturesque qualities are open for public consumption with little threat of being worn down from that sensorial consumption. At least as compared to the consumption of its foods, fibres and energy (the first two columns of the Table of the Checklist), and with its "consumption" (non-availability) through owner-ship or its users' rights through lease or tenancy contracts (columns 3 and 4 of the Table of the Checklist). But although this section 3.2.3 focuses on the least material, qualitative aspects of the landscape, the landscape itself is always also material as elaborated on in section 3.2.1 "Criteria for the (a)biotic realm (environment and ecology)". Thus, both sides of the landscape have to figure in the definitions used here. Several definitions will be mentioned to show how they all, from different point of view, grab important parts of the whole.

All peoples' histories and therein all human biographies are embedded in (combinations of) landscapes: the sea, the mountains, the desert, the steppe, the forest, the delta, the riverside, the small or the large towns cities and mega-poles (Baldock and Beaufoy, 1993; Giorgis, 1995; Schama, 1995). "Landscape is seen as a characteristic spatial arrangement of land-units" (Vos and Fresco, 1994). Therein the landscape definitions surpasses the realm of natural sciences, as it refers to the *characteristic arrangement* of the units (species, bio-topes, infrastructure, buildings), that is, it refers to an artistic, aesthetic notion addressing the vision of the whole. It is that vision of the whole, that perception of the pictorial quality that makes the landscape into a landscape. The landscape as a whole delivers the context wherein the mentioned elements of the nature, which can be analysed by natural sciences and used for design by technical sciences, do figure. Those landscape elements of nature and culture must be discerned within and reduced from the landscape as a whole, before they can be analysed in one way or another (Bockemühl, 1992).

In other words: "The landscape is a physical environment of objective measurable attributes. Apart from these physical features, landscape has inter-subjective qualities perceived and valued by people, qualities that depend on the characteristics of the observer. And as cultural and aesthetic values of observers change over time, the images and values of landscape change over time too. Therefore, it makes no sense to consider landscape as a static phenomenon that should definitely be conserved in a certain fixed state"' (Vos and Fresco, 1994). The last part of this statement encourages the thinking and valuing in terms of sustainable development of landscape, which is the key issue of the approach presented in this checklist.

In this checklist of criteria for the development of sustainable rural landscapes, criteria for the cultural environment are included, because the landscape importantly influences the welfare and well being of people and their ability to survive in a society at large. Vice versa, a culture that appreciates and cherishes its landscape values, implementing that appreciation into a well-structured social life and a healthy economy, is a prerequisite for a sustainably managed landscape. To achieve an appreciated landscape, Fairbrother (1974) mentions that "Landscape is essentially the physical expression of land use and it is with (good) land use that we have to begin".

From the seventies of this century onwards, a strong renewed interest for the landscape arose with, as a new trend, concern for the re-association of the different meanings that landscapes imply. The justification of all the specific perceptions and reflections of those involved in the landscape, is certainly still one of the greatest of present challenges for the landscape research (Terrasson and Le Floch, 1997). In other words, this same challenge is addressed as that of orchestrating the multifunctionality of the landscape. In that context, the Dutch government refers to the three E's: Ecology, Economy and Ethics as leading themes.

The multiple landscape functions, each and all going along with their specific reflections and value-systems, can be indicated as follows. The landscape is the source of our soil, air and drinking water. It is also a place to produce, to recreate, recover and revalidate from work, stress and disease, to do research, to get inspired for works of art, to enjoy all features of nature and its species, and altogether to develop one's individual personality, together with others and alone. The challenge of post-modern society is indeed to find ways to orchestrate these multiple purposes of the landscapes, on a relevant set of scales, in such ways that the functioning of one landscape part does not affect the other parts in the detrimental way it now so often does. Therefore, compatibility of landscape uses is required and a sustainable approach of land-use, as referred to in this study (see column 1, environment, and column 2, ecology, of Table 3.1), is seen as most feasible. The detrimental effects of current rational land-use on many of such rural landscape qualities, which are neither easily quantifiable nor easy to sell, will be discussed later on.

Referring to the issue of appreciation and aesthetics, nowadays citizens, somehow active in all kinds of nature conservation and environmental care oriented action groups, criticise a domination of the agricultural production approach that creates so much monotony in landscapes to the detriment of most other activities mentioned. So there is a discrepancy between attitude and behaviour, seen on the level of society as a whole (Yi-Fu Tuan, 1972). On the one hand there are the farmers, altogether less than 5% of European professionals, but nevertheless using up to 80% of the European landscape. Thus the farmers are the main consumers of the rural landscape with their property rights and obligations. On the other hand, there are many other groups, all with different attitudes, that want to use and appreciate the landscape for their particular set of objectives, but with less clearly defined rights and less defined obligations (see column 4, sociology, of Table 3.1). The citizens demand the lowest possible food and fibre prices, whereas on the other hand they demand for landscape, as mentioned above, often without being aware of that discrepancy at all. "But whatever the urban accessories, our longing for the country is real: it is a genuine desire for the natural rather than the man-made" (Fairbrother, 1974). The urban citizens like for example to see and hear many kinds of flowers, trees, birds, butterflies and a variety of (semi) wild mammals. In most farmers' perceptions all these are enemies of the volumes they produce for society, according to its demand as expressed in the fact that only food and fibre volumes are paid for. The mentioned demand for landscape quality has as yet found little to no hold in appropriate payments. As a secondary effect of the successful increase of agricultural volume production, the percentage of farmers diminishes rather fast. Thus, ironically, the profession that originally created most of the appreciated landscape, organised itself out by breaking down the landscape for the presumed benefit of the citizens (society) who do not pay for anything else than food volume. The landscape itself will get influenced more and more by the ideas and concepts from non-farmer groups, which so far left the farmers unpaid for their service to create an appreciated landscape. Non-farmer groups now explicitly demand for this unpaid service of farmers at the moment that the original producers (farmers) of the landscape values have almost left.

In broad circles of European society and authority, restoration of the ecological quality of the landscape is seen as a step further in the cultural development. Organic types of agriculture may play an important role in this restoration process, as they represent a feasible style of farming based in a concept of respect for and an attitude of co-operation with nature and the environment. In general, they already provide a clearly richer type of landscape with more naturalness than non-organic (conventional) farming (Colmenaris and De Miguel, 1997; Hendriks et al., 1997; Van Manvelt and Mulder, 1993).

Amongst the landscape problems at stake, the following major categories of detrimental effects can be indicated:

1 Desertification or standardisation of the landscape, as its regional diversity decreases together with the vanishing of many natural landscape elements from the rural area (bio-topes of various sizes);
2 Fragmentation of the landscapes, especially in the neighbourhood of urban areas, took place as new infrastructure lines were recklessly forced upon the existing ones;
3 Simplification of the landscape, as too many extensive and abrupt changes have too often been taken place, disturbing the appropriate development of landscape elements, making mature landscapes that show the full potential of their features and qualities rather rare;
4 Ongoing and alarming decrease of bio-diversity, along with the before mentioned trends.

Under the leading paradigm of rigorous quantification, commoditisation and modernisation of land reform and agriculture have replaced pre-existing land use systems and related landscape elements (e.g. hedgerows and single trees in fields, drinking ponds, old drainage and irrigation systems disappear). The bio-mass production has won at the costs of the cultural landscapes, that were either destroyed or just vanished (Baldock and Beafoy, 1993; Giorgis, 1995; Vos and Stortelder, 1992).

Both, in the case of extensification and intensification, cultural landscapes are threatened. Large-scale agricultural landscapes with an increasing intensity of external input land use are called 'stressed landscapes', which usually are characterised by monotonous or pattern-less uniformity (Vos and Stortelder, 1992). New elements, replacing the former diversity, are not concordant with the pre-existing spatial arrangements: they tend to dominate entire areas, like for example widespread mono-cropping of bulk products: ryegrass, maize, small grains, olive groves or vineyards. Or the new elements are dispersed without any local physiographic differentiation, like for example isolated stands of Eucalyptus and coniferous trees. Also many modern agro-industrial buildings can be quite disturbing in the landscape, such as silos for feed, batteries for animal production, storehouses, sheds and silos. Here however, not only the modern technology plays its role, but also the land-ownership dispersal, mainly due to heritage related effects, and building permits together with regional planning procedures and standards.

For the assessment of the human-science aspects of European landscapes, we again propose to use two related sub-realms. The first sub-realm refers primarily to the appreciation of rural landscape by its users and inhabitants: column 5 (psychology) of the Table of the Checklist. The psychology realm is based on non-expert values, which are largely derived from landscape psychology that systematically identified aspects of the landscape which its users observe and appreciate or depreciate (Coeterier, 1996). The psychological criteria and parameters are meant as a checklist for all respondents' perceptions of the landscape at stake.

The answers per respondent may differ quite a lot. However, there are also common denominators, shared visions on appreciated landscape aspects that are important starting points for a well-calibrated decision-making. The second sub-realm refers to expert validation, mainly derived from physiognomy, geography and landscape architecture: column 6 (physiognomy and cultural geography) the Table of the Checklist), and drawing on the landscape's cultural history and its imaginable future. The expert is presumed to know about the cultural history of each landscape unit at stake, for instance by using old maps, by reading the features of the surrounding landscape of the farm and by imagining an idea of an 'ideal landscape' under the chosen conditions (Colquhoun, 1997; Kuiper and Paans, 1990; Vereijken et al., 1997). For the psychological criteria experts are needed to ask the non-experts the right questions on their expectations and appreciation. For the physiognomic and cultural criteria experts are needed to co-operate with landscape users to make sure that local knowledge and feasible imaginations are brought up, complying to the perceptions and demands of those affected by the landscape changes.

The criteria for the assessment of the cultural aspects of the landscape draw heavily on the physical survival criteria for the a-biotic and the biotic environment (column 1 and column 2 of the Table 3.1). The implementation depends on the decision-making procedures of the social environment, in view of the economy as perceived by the actors in charge (column 3 and column 4 of Table 3.1).

Quality of the cultural environment: Criteria for the humanity realm	
5. Psychology Subjective regional landscape appreciation	**7. Physiognomy /** **Cultural geography** Objective regional landscape identity
5.1 Compliance to the natural environment	6.1 Diversity of landscape components
5.2 Good use of the landscape's potential utility	6.2 Coherence among landscape elements
5.3 Presence of naturalness	6.3 Continuity of land-use and spatial arrangement
5.4 A rich and fair offer of sensory qualities, such as colours, smells and sounds	
5.5 Experiences of unity, like for example: completeness, wholeness and spaciousness	
5.6 Experienced historicity	
5.7 Presence of cyclical developments, for example growth cycles and the seasons Careful management of the landscape, for example at the level of maintenance	

5 PSYCHOLOGY (Column 5)

TARGET: Appreciation of the rural landscapes as spontaneously experienced by all participants is important for the participating population as well as for the landscape's sustainable development. For all participants, farmers, inhabitants, tourists and people who need to recover from stress, mental or physical diseases, there should be places and possibilities in the landscape to feel comfortable, at home, secure, inspired and empowered to recreate, relax, recover, or otherwise be efficiently engaged in professional and leisure activities. The landscape should be inviting and inspiring for people to dedicate their labour and leisure actions in favour of the landscape and its users. Therefore, landscape's multitude and ranges of qualities should provide sufficient interesting information and pleasure.

From sociological research (Volker, 1997), it has become clear that the public's landscape appreciation is not "only" important for the public's wellbeing, but also for the sustainable management of that landscape. People are willing to contribute to the landscape's management, be it financially or by investing private time and energy, but only insofar they feel appreciation, pleasure and compassion for that landscape. If they do not like it and identify, they are not motivated to get involved in and take responsibility for its management.

> **Illustrations to the importance of qualities for public wellbeing**
>
> "I think the sensory quality must be considered in planning for an entire inhabited region, and for the everyday environment, experienced in the whole range of daily actions. Oppressive smog, monotonous noise and heavy odours are all restrictions on our sensing. Removing these restrictions is the precondition of all purposes to be described. Extreme sensory conditions like ear-splitting noise, intense glare and prostrating heat can prevent us from functioning or cause organic damage. More often, sensory conditions disturb our comfort or reduce our efficiency" (Lynch, 1976).
>
> "We can hear sounds that tell us about the size of the spaces, the surface of the walls, the location of the sounds, the rapidity of movement from others and ourselves. We can therefore orient ourselves, which adds some feeling of security" (Vroom, 1986).

This illustrates the link between the sociological criteria (column 4 of Table 3.1) and the psychological criteria (column 5 of Table 3.1).

Coeterier (1996), through interviewing inhabitants in a research that covered a period of twenty years, collected, inventoried and systemised a wide range of criteria that reflect the public's landscape appreciation. Only when these criteria are known, they can be included in discussions and decision-making. The clearer the various kinds of appreciation and their distribution over those involved in and affected by the landscape planning have been made explicit in the dialogue with actors, stakeholders and experts the more 'as-well-as' or 'win-win' solutions become feasible.

The kind of impressions people get in a landscape can be summarised as: colours, sounds, smells, tastes, humidity, temperature, wind, light and shadow. They affect the public in two ways: (1) through sensations, primarily addressing the empathic realm of feelings, and (2) through information, as messages from the landscape's identity, more addressing the realm of cognition.

Pohl (1995) describes the Wenzel's Cosmic Park concept, wherein all senses contribute to perceive nature as a whole. Direct contact with natural features like water, temperature, smell and earth, should provide multi-sense perceptions and a feeling of freedom from all day life. The sounds of rain drops on Lotus leaves and of wind through Bamboo, the sight of the season's different colours, were leading themes in the creation of old Chinese gardens (already 3000 years ago) and parks (known from the 9th century). The scholars wrote about these experiences in poems (Hongxun, 1982).

Eight main criteria have been distinguished with respect to the appreciation of landscape.

5.1	Compliance to the natural environment
5.2	Good use of the landscape's potential utility
5.3	Presence of naturalness
5.4	A rich and fair offer of sensory qualities, such as colours, smells and sounds
5.5	Experiences of unity, like for example: completeness, wholeness and spaciousness
5.6	Experienced historicity
5.7	Presence of cyclical developments, for example growth cycles and the seasons
5.8	Careful management of the landscape, for example expressed as the level of maintenance

5.1 Main criterion: Compliance to the natural environment

MAIN TARGET: To warrant that the a-biotic features, like the geomorphology, relief, soil and water, which are well respected as carriers of the landscape and as basic conditions for all kinds of land-uses, actions and perceptions. This compliance should be visible and accessible for public experience.

Parameter for the compliance to the natural environment:

1   Clear presence and cultivation (conservation) of the region's special natural features like water-bodies of all sorts, slopes, peaks, marshes, dunes and cliffs.

5.2   Main criterion: Good use of the landscape's potential utility

MAIN TARGET: The obvious visibility of a legitimate use of all functional aspects of the landscape, according to society's demands and the region's carrying capacity. Good use is considered as a main attribute of the good landscape. For farmers the rationality of the use and the way rationality is shown is of crucial importance for their professional self-esteem. Areas that are too intensively used exclude or deploy all feeling of naturalness. Respect for nature's intrinsic values does very well show-off in such types of management that favour nature's processes and the effects on mankind. Opportunities for various types of private enterprises (production) should be available, presumed they comply with the criteria mentioned before in the other columns of Table 3.1 (environment, ecology, economy, sociology). Also opportunities for leisure and recuperative activities like biking, riding, walking, tracking, climbing, swimming, fishing or watching the clouds or the rivers flow, are important for inhabitants and tourists alike. However, a transparent dialogue is needed to make sure that opportunities for leisure and recuperative activities are implemented in a way that is compatible with the landscape's character, appropriately located and restricted to a sustainable intensity of use.

Parameters for a good use of the landscape's potential utility:

1   Rationality of the sustainable land-use and the way it looks like. Good used of the land should be visible.
2   Percentage of sustainable areas in proportion to the whole landscape and those managed in unsustainable ways
3   Possibilities for activities other than food and fibre production, on their feasible locations and their appropriate intensity of actual use

5.3   Main criterion: Presence of naturalness

MAIN TARGET: To respect the natural heritage in its intrinsic values, warranting the perceived natural wellbeing of the bio-sphere and its components. The way in which a region's inhabitants perceive naturalness may differ from that of citizens and from expert's opinions. The perception of naturalness also depends on trends and awareness in society. At the moment, societies' demands for strong feelings of naturalness are greatest in the European countries with the most external input and intensive agriculture. Education of the awareness for nature's qualities and values is of paramount importance for the sustainable management of the landscape.

Parameters for the presence of naturalness:

1   Indications that the landscape has developed in a sufficiently natural way
2   Dominance of natural elements, lines, patterns, materials, over artificial ones
3   Presence of natural, non-productive sites and old trees

5.4 Main criterion: A rich and fair offer of sensory qualities

MAIN TARGET: To warrant a wide and diverse range of appreciated, recuperative and inspiring sensory qualities like smells, sounds and visual elements, under absence of notoriously intrusive and stressing impressions.

Parameters for a rich and fair offer of sensory qualities:

1   Smells
2   Sounds
3   Visual perceptions
4   Spatial perceptions

1   Smells

This parameter is assessed through (1) the presence of smells from healthy crops, flowers, herbs, soil, compost, trees and leaves or (2) the absence of repellent or nauseating smells, for example from bio-industry or slurry.

2   Sounds

This parameter is assessed through (1) the presence of appreciated nice sounds of mammals, birds, insects, streaming water, and wind or (2) the absence of unfavourable noises from machines, motorways, industry and airports.

3   Visual perceptions

This parameter is assessed through (1) the presence of sun, half-shade and shade, various natural colours, textures, contrasts, grades and gradients or (2) the absence of unidentifiable, displaced and non-fitting impressions.

4   *Spatial perceptions*

This parameter is assessed through (1) the protected places to hide from extreme weather conditions (for humans and animals) or (2) the open places to feel exposed to the weather, meeting the forces of nature (id.).

5.5 Main criterion: Experiences of unity

MAIN TARGET: To warrant the experience of completeness, wholeness, unity and spaciousness of the landscape. Unity here refers to the landscape as a whole, with all parts fitting well together and functioning as a healthy living organism, with a character and identity of its own. Unity refers to an added value, that none of the single parts possess. In that particular perspective, the character of the whole is more important than the parts or any of the parts.

Parameters of the experiences of unity:

1   Order
2   Completeness
3   Wholeness
4   Spaciousness

1   *Order*

This parameter is assessed through the presence of unity and order. Here, in classical terms, it is the subtle and playful balance between Appolonian order and Dyonisian chaos or the swaying balance between death and life that counts. (Van Mansvelt, 1997; Schiller, 1981). The aesthetic feeling for harmony, character and identity that obviously needs ongoing education and discussion are most directly addressed in this parameter. Some well-placed dissonance can for example be part of an appreciated character or piece of art.

2   *Completeness*

This parameter is assessed through the presence of all the appropriate elements and thus completeness of a landscape. Here, the issue is to clarify which elements should be present, on what scale and to what extend, according to the characteristic features of the landscape at stake, as seen in the perspective of a sustainable development that complies to its identity.

3   *Wholeness*

This parameter is assessed through the absence of non-fitting and thus disturbing elements. Here, the issue is to come to an agreement about which elements do too much disturb the landscape's character in perspective of a sustainable landscape development and which do comply with the landscape identity.

4   *Spaciousness*

The appreciation of spaciousness partly depends on the type of landscape and education (history) of the observer. In a wide and open landscape some trees, for example along a road or around a farm, may accentuate the open character, whereas too many trees may obstruct the views. In a small-scale landscape, the removal of all hedges and woodlots is experienced as detrimental to its quality of naturalness. However, a few views, down lanes or through some open fields, may accentuate the landscape's intimacy.

5.6   Main criterion: Experienced historicity

MAIN TARGET: To respect the cultural heritage and to warrant the presence of the site or landscape's historical character, thus allowing for the perception of that cultural heritage. Just as the active human memory is crucial for the development of its identity, so does the landscape's identity depends on the perceivable presence of its history. Like for other issues mentioned before, does the assessment rely on appropriate education at primary and secondary school and other types of education (courses, training, etc.). Awareness of history is an indispensable prerequisite of any culture, agri-culture fully included.

Parameters for the experienced historicity:

    1   Historic elements of art and crafts
    2   Historic landscapes patterns

1   *Historic elements*

This parameter is assessed through the presence of old elements of art and crafts (industry) like farmhouses, churches, mills, sanctuaries, castles, irrigation systems, bridges. Especially when reminders of various historic phases are present, historicity is well served.

2   *Historic landscape patterns*

This parameter is assessed through the presence of old landscape patterns, for example from roads, fields, canals, rivers, plantings and terraces.

5.7 Main criterion: Presence of cyclical development

MAIN TARGET: To warrant the presence of sensorial attributes of the subsequent seasons. The change of the seasons presents naturalness in a very strong way. It shows how the wheel of life keeps on turning, reminding the observer of humans' own development, as a metaphor based in nature's reality. The rhythmical change of seasons, which is never the same and always varying on the old familiar pattern, communicates with the feeling of health. It always balances between youthfully expanding growth and ripening contraction in fruit-building, seeding and later decay. Both latter processes precede the next season's or next year's restart, resumption or recycling of the development.

Parameters for the presence of cyclical development:

1	Developmental phases of natural elements
2	Landscape maintenance cycles
3	Succession of landscape bio-topes
4	Decomposition

1   *Developmental phases of natural elements*

This parameter is assessed through the presence and visibility of all phases of the natural elements' developmental phases, fitting the seasons' particular qualities as expressed in colours, smells, moisture, temperature, light and life's activities in general.

2   *Landscape maintenance cycles*

This parameter is assessed through the presence and visibility of landscape maintenance cycles that comply to the daytime, season or year (age) of the landscape and its elements.

3   *Succession of landscape bio-topes*

Evidence of appropriate succession of the landscape's bio-topes, warranting that not only young pioneering ones but also riper ones, with a climax character are present (see also completeness).

4   *Decomposition*

Evidence of appropriate decomposition. This refers for example to the presence of old, dying and dead trees in forests or plantings, but also to the presence of appropriate places for composting, manure storage and manure / compost application.

5.8   Main criterion: Careful management of the landscape

MAIN TARGET: Careful management of the landscape expressed as good maintenance or upkeep of the cultivated landscape elements to show that the land is well kept and cared for. Now this 'good' or 'no-good' is a question, which needs clarification for those involved. On the one hand, there is a 'nice-and-tidy' trend, demanding that any crop in a field looks equal and that no other plants, called weed, are present at the field. On the other hand, there is a demand for naturalness requiring a certain amount of variation and diversity within species and of species and bio-topes. As modern agriculture goes along with a strong mechanised type of order, it tolerates less natural variation then traditional land-use systems did. However, this modern order does at least partly reflect over-management, whereby coherence in time and space can get lost. For the sustainable management of the rural landscape, compliant to the criteria mentioned before, appropriate norms for neatness and good management must be developed.

Parameter for a careful management of the landscape:

1 Farm succession

1   *Farm succession*

Evidence for good management is balancing between such extremes as over-management and neglect and between efficiency and over-exploitation. This should for example comply with the original farmers' ethics: to leave the successor(s) a better farm (soils, seeds and stock) than one inherited from the predecessor(s).

6 PHYSIOGNOMY/ CULTURALGEOGRAPHY (Column 6)

In addition to the approach of column 5 of the Table of the Checklist (psychology), which focuses on the experiences and perceptions of all landscape stakeholders who somehow use the landscape for their consumption and/or production, this column 6 of the Table of the Checklist (physiognomy and cultural geography) focuses at the available knowledge and conscious awareness of all issues at stake. In column 6 of the Table of the Checklist (Table 3.1), concepts, visions and approaches are most explicitly discussed, while the previous columns of the Table of the Checklist, particular column 1 (environment), column 2 (ecology) and column 3 (economy) are more oriented around facts and figures, presuming that the empirical data are "free of value". We presume that, within the objective of inter-disciplinary synergy between co-operating expert teams in charge of landscape and land-use planning according to sustainable management, facts and figures will be successfully put in the context and concepts will be successfully implemented. Thus, all relevant aspects as presented in column 1 till 6 of the Table of the Checklist (Table 3.1), can be appropriately considered and effectively taken into account.

Lynch (1960) mentioned *identity, legibility and recognition* as important criteria for the assessment of the aesthetic landscape quality. He stresses that landscape features, urban or rural, are necessary to allow the inhabitants and other landscape users to orient themselves on their necessary feelings of being home and welcome instead of feelings of strangeness and being lost. However, such an orientation can not always be expected everywhere and from everyone at first sight. A certain time to acclimatise, to land and to find ones way is definitely needed in most landscapes for those to whom the landscape is new. Readiness to undertake activities to get to know and understand the landscape can be seen as a fair request towards the observer or new stakeholder of the landscape. At the other side, the landscape character as a pattern of high continuity with many distinctive and clearly interconnected parts should communicate in such a way that it is being worth to put effort in getting to know and understand the landscape (Lynch, 1960). Vroom (1986) also mentioned that landscapes must be identifiable and thus allowing for identification and explanation of its structure and its specific elements.

Orientation in space and time is the framework for recognition. "The identification of places, as well as their organisation into mental structures not only allows people to function effectively, but is also a source of emotional security, pleasure and understanding. It is the rate of sensible changes as well as their direction and connection with past and future that has the strongest impact on people. No place remains unchanged. People are pleased to know or to understand its history" Lynch (1976). This quotation of Lynch again refers to the importance of inhabitants' and users' sympathy for and commitment to a landscape to warrant its sustainable management as based on a fair amount of knowledge and serious interest.

Furthermore, many theories on landscape ecology have been published, discussing the best physical features and spatial arrangements for species diversity, connectivity between ecosystems (coherence) and persistence of ecosystems (continuity). See for instance, Forman and Godron (1986) and Soulé and Simberloff (1986). Coherence and continuity have already been discussed in section 3.2.2 "Criteria of the social realm: economy and sociology", but are mentioned here again because of their aesthetic impacts. Within both disciplines, architecture (column 6 of Table 3.1) as well as ecology (column 2 of Table 3.1), the vertical and horizontal spatial and temporal relationships in the landscape are perceived as very important and should be analysed for their coherence. Together they cover the so-called landscape structure. A further elaboration of the ideas about the vertical, horizontal and temporal relationships in landscape plans at different scale levels have been analysed by Kuiper (1998) in a quality assessment of landscape planning in the Dutch river basin. In another study of Kuiper (1997), the main criteria, diversity, coherence and continuity, are used as a common basis to plan objectives with special regard to aesthetic and ecological quality of organic agriculture in a brook valley.

To formulate general standards for the parameters of the main criteria (diversity, coherence, continuity) for the quality of the cultural environment is not without problems and even impossible. Therefore it is disputable to make EU regulations out of these criteria. For example, the perception of naturalness depends not only on the observer's biography, but also on the landscape at stake. The ideas about nature of most urban citizens are quite different from the ideas of farmers. Nature conservationists from the city may be more radical than the ideas of farmers who are more familiar with that particular rural area. Moreover, the appreciation of any landscape feature, like spaciousness, coherence or whatever, depends on the type of landscape these features figure in (Coeterier, 1996). Fairbrother (1974) mentions that "our reactions to sun and shade, coolness and warmth, and so on equally depend on the context".

Aesthetic demands for any length or surface of hedges per area or for a particular amount of trees and woodlots are not adequate for all Dutch or European landscapes. However, as pointed out in column 2 (ecology) of the Table of the Checklist, there are clear indications that certain types of faunal or floral species require a habitat of a certain size. So the aesthetic discussion can be served by adding ecological evidence and vice versa. The ecological discussion can be served by referring to historical evidence and aesthetic demands of the relevant groups of society. In this study, we state that for a fair and compatible set of regulations for European landscape management, including agro-sypvi-pastural management, a comprehensible set of criteria and parameters should be present to serve the discussion and decision-making of those in charge. By taking into account all points mentioned and explicitly deciding on their proportional priority, the chance that sustainable management is served does most presumably increase, see chapter 5 "Uses and users".

TARGET: The landscape should reflect its characteristic natural and cultural heritages and also its present use and meaning. Its a-biotic features should be respected, together with its biosphere, as a source of information about the landscape fundaments. Landscape patterns out of the past should be respected and possibly used in the landscape functions of today. Thus the landscape should reflect its past and present uses and meanings to supply information that is of interest and to satisfy the needs of nowadays society, without affecting the needs of future generations.

Note: The main criteria, diversity, coherence and continuity, together are demands of experts in order of landscape quality. The landscape should allow observers to identify the different landscape components, to explain the order and relationships between landscape components and to recognise the cultural and natural heritage. The landscape should allow for orientation in space and time. The three main criteria are interrelated and inseparable. For the criteria used in this section of the checklist, the methods for landscape analyses from ecology and landscape architecture have been consulted from Lynch (1960, 1976) and Vroom (1986).

Three main criteria have been distinguished:

6.1   Diversity of landscape components
6.2   Coherence among landscape elements
6.3   Continuity of land use and spatial arrangement

6.1   Main criterion: Diversity of landscape components

MAIN TARGET: Satisfying the various demands of society for landscape functions in such a way that it supplies the relevant information (legibility) without stressing the landscape's sustainability. The diversity should reflect the relationships between land use and the landscape's a-biotic features as the fundaments of the landscape's identity. At the other side, diversity should be restricted by its characteristic historical and social development and through its a-biotic features. Extreme diversity of landscape elements can lead to a visual and ecological chaos with far too little coherence in space and time. A maximum kind of species diversity present in zoos or megapoles is not preferable at landscape level.

Note: Diversity is a complicated criterion with a need for scale specification (De Jong, 1978). As pointed out by Kuiper (1997), a species or bio-tope-concentration model shows homogeneity at local level but diversity at regional level, while the even-sprawl model has the inverse effect. The conclusion is that landscape diversity depends on the scale level under consideration, ranging from farm-level to European landscape.

Parameters for the diversity of landscape components:

1	Diversity of landscape types per country
2	Diversity of landscape units (bio-topes) per landscape type
3	Diversity of elements (crops and planting) per landscape unit
4	Diversity of species per bio-tope

6.2   Main criterion: Coherence of the landscape elements

MAIN TARGET: In favour of the legibility of the landscape, landscape should reflect its basis in geomorphology (vertical coherence), the interconnectedness of its elements in order and structure as a whole (horizontal coherence), and the development phase of elements and the time of the days and year (temporal coherence). The coherence of the landscape should facilitate people to orient themselves in time and space, by feeding people's physical maps of the rural environment. Moreover, as already discussed in column 1 (environment) and column 2 (ecology), the connectivity of the landscape serves the necessary connectivity for the survival of the characteristic species and bio-topes.

Example about the coherence of the landscape components

Visible hydrological systems and road patterns provide necessary information for the farmer, the inhabitant and the tourists. The difference between main roads, secondary roads, land-roads or death-end roads is important for drivers. Road systems should improve the landscape's accessibility, but also not harm too much the ecological quality and the landscape character. Similarly, it can be of interest to see if the watercourse is a ditch, a brook or a river and how far the flooding can come.

Parameters for the coherence of the landscape components:

1	Hydrology
2	Infrastructure
3	Farming
4	Ecology

1   Hydrology

The appearance of the watercourses should reflect their place in the order of the hydrological system as well as the water quality and the season's character.

2   *Infrastructure*

The appearance of the roads should reflect their place in the system and should improve the accessibility of the landscape without harming the landscape's character.

3   *Farming*

The spatial arrangement and temporal use of fields and the farms should reflect the locally adapted and sustainable way they are managed like for example organic farms tend to do.

4   *Ecology*

The plantings or natural sites added to the farm should contribute to the ecological network for connectivity, bio-diversity and seasons compliance, within the context of legibility of the region's historical development.

6.3   Main criterion: Continuity of land use and spatial arrangement

MAIN TARGET: To warrant the legibility of the landscape as an expression of time and different periods of cultural history. The rate and extent of the changes in the landscape should not exceed people's mental grasp. If changes in land-use and spatial arrangement are too fast and abrupt then the characteristic features of the landscape, like species belonging to a farming system, will disappear. Subsequent old maps of the site at stake show the landscape changes on paper. Recent changes result in a young landscape with young trees and vegetation only, which may look like the results after a big crash and will make people feel uncomfortable. A landscape is more interesting and rewarding if more time dimensions can be perceived.

Parameters for the continuity:

1   Cultural history
2   Duration and continuity of land use and spatial arrangement
3   Presumed future sustainability

1   *Cultural history*

The parameter is assessed through the presence of elements and patterns that remind of special periods in cultural history.

3   *Presumed future sustainability*

The presumed future sustainability is assessed through the expectations derived from the present management in the context of the land's potential use and in particular with regard to the health of the soil and the socio-economic developments.

3.3   CONCLUSIONS

The synergy or compatibility of the targets of the criteria in the Table of the Checklist

Theoretically, the position is chosen that similarly as in the interaction of human motivations are leading human's life. So do the requirements for a sustainable management of the landscape inter-act within the landscape's historical development. Though there might be a local and or temporary focus on one or more particular targets, leaving others under focused for some time, through cyclical processes, rotational shifts of locations and varying scales, polar tendencies can be merged in an over-all development. The more options for 'as-well-as' concepts are chosen instead of concepts of an 'either or' nature, the more synergy can be detected in the landscape's multi-functionality. Taking the extremely fine-tuned interactions of organs with opposed functions in an organism's body or the 'competing' species' interactions in ecosystems as examples, it is the challenge for those in charge of the landscape's management to find how the similarly opposed functions can be merged into a synergetic system. Presuming mankind's co-evolutionary evolution, the position can be chosen that on the long run mankind and nature have compatible interests, supporting the concept of a common future.

Finally, the following concluding remarks can be made about the checklist for sustainable landscape management:

1   The above mentioned parameters of the checklist are meant to facilitate the realisation of the qualitative targets defined per issue. These targets are meant to be instrumental for the realisation of a landscape management of rural and agro-silvi-pastural areas that warrants a sustainable development for mankind and nature for next centuries. In the sense of Maslow's concept, the primary need for ongoing development is the motivation of humans to make sure that the biosphere is functioning good enough to keep mankind healthily nourished with water and food and sheltered with clothes, houses and heat. However, in relation to the social and cultural demands, which will be considered in the next columns, it is important to realise that over-consumption of food in general and over-consumption of animal protein in particular, burdens very heavily on the global available resources. The same holds for the over-consumption of fossil, non-renewable energy and other limited resources. The more a society manages to shift from maximal tolerable consumption levels to minimal required consumption levels of all limited resources, especially in the monetary rich countries, the better the perspectives are for a sustainable development of landscape as the basis for human livelihood.

2   For all quantitative parameters it is important to be aware of their source in normative validation of the relevant facts and interactions. Numbers of species per surface unit, amounts of nutrients per volume of feed, food, ground water,

surface water or manure: they all have their meaning only and alone within a wide context of adjacent considerations and namely intentions (goals, targets). As part of a culture, paradigm, set of values, habits, or appreciation, the so-called hard facts and figures figure in their context. Also, the perception of causality is inevitably defined by the context of a discipline and is part of a paradigm. So, although quantification and hard facts are needed, the context in which they figure and the norms they inevitable incorporate must be included in the dialogue in which the hard facts are used. Beside the need to specify means to reach objectives and the related parameters to assess the degree of success in reaching those objectives, the tendency to make goals out of the parameters must be vividly kept in mind as misleading. So, here it is strongly recommended to make the objectives to be reached by any measurement a constant issue of all dialogues of landscape management.

3   With both previous points is mind, it must emphasised that this checklist, as offered to any group and for any purpose (see chapter 5 "Uses and Users"), should be used to support and widen the dialogue between those in charge of whatever kind or level of landscape management. In the spirit of the concerted action underlying this checklist, it makes only sense if this checklist is used in an interdisciplinary and co-operative team where local experts and specialists on all fields at stake do participate. From our experiences and the considerations of next point, it seems very worth-wile to make sure that farmers who are familiar with all ins and outs of organic farming are invited to participate in such teams.

4   For those familiar with the research, theory and practices of organic agriculture, it will be clear that most of the recommendations made, most targets set and most parameters proposed in order to warrant a sustainable landscape management, are fully in line with those of organic agriculture. This makes no wonder, as the compatibility of farming practices with a long term fertile soil, healthy crops and husbandry as well as a healthy landscape and environment are at the roots of all versions or types of organic agriculture (Van Mansvelt and Mulder, 1993). This statement will be elaborated in more depth in chapter 4.

CHAPTER 4   PERFORMANCES OF ORGANIC AGRICULTURE

4.1   THEORY AND LITERATURE

Within the framework of agricultural policy making, the world wide demands for sustainable development in general are focusing on the sustainable development of the rural areas in particular (FAO's SARD and UNCED's Agenda 21 /Ch 14). In the definitions of sustainability, four fundamental aspects are mentioned: (1) food security; (2) employment and income generation; (3) environmental or natural resource conservation; (4) people's participation and empowerment. In different perceptions different priorities may be set in their implementation, expressing different attitudes toward nature, society and the ethical decisions involved. In line with these differences, different strategies for agricultural developments are favoured, each with specific consequences for the development of rural landscapes.

Different definitions and different concepts of agriculture are reflected in different implications for farming in practice as well as for the design of farming systems (Altieri 1989; FAO 1992; IFOAM 1996; Neher 1992; Van Mansvelt and Mulder, 1993; Schultink 1992; Vereijken 1992). The most fundamental issues, shared by various perceptions of sustainable agriculture, are present in the main goals phrased by the FAO (1992), see Table 1. These main goals regard the basic motives of humans as ranging from food security to empowerment and self-realisation and do comply with the human motivations as phrased by Maslow (1968).

For a world wide implementation of the FAO-SARD / UN Agenda 21 for agriculture and EU policy for the European land-uses and landscapes, feasible specification of program objectives into appropriate schemes for planning and evaluation are required. For every farming sub-system special objectives can be formulated, eventually leading to 'checkpoints' at farm level, land-use level and rural landscape level. So for example Altieri (1989), FAO (1991), Schultink (1992), Van Mansvelt (1992), Vereyken (1992) and others formulated more specific objectives and criteria for comparison of farming systems in regard to their sustainability. Using such sources, more specific objectives derived from FAO's values and interests are listed in the second column of Table 4.1.

Table 4.1: Objectives for sustainable rural development

Basic values and interests of sustainability, as phrased by FAO (1992). *Human motivations*	Specific objectives of sustainability, as mentioned by: Altieri (1989), FAO (1992) Van Mansvelt (1992), Schultink (1992).
Food security. *Physical survival*	Self-sufficient supply of good quality food, fibre and (renewable) energy.
Employment and income generation in rural areas. *Social survival*	Diversification of income, Labour demand in rural areas, Social security, Socially and culturally acceptable technology.
Natural resource conservation and environmental protection. *Earth's survival*	Bio-diversity, Regenerative potential of nature, Use of local resources, Health and well-being of the ecosystem, Local landscape values.
People's participation and human resource development. *Ethical survival*	Human resource development, Self-promoting and self-help potential, Farmer's and community's empowerment.

The choice for the roads that will lead to these goals is greatly influenced by the values, expressed as priorities, that policy makers give to each of these issues (Devall and Session, 1985; Verhoog, 1980). Generally, High External Input Agriculture (HEIA) and Integrated Agriculture (IA) give priority to food production, whereas Low External Input Sustainable Agriculture (LEISA) and Organic types of Agriculture (OA) try to give more balanced attention to all values of multi-functional land-use. This expresses an integrative approach of the latter post-modern movements, which commit to an integrative approach ("liga et serve – relate and server"), whereas the former modern movements strongly believe in the power of a segregation ("divide et impera – divide and rule") (Van Mansvelt and Van Elzakker, 1994). To support awareness on the impact of both attitudes on the agro-sylvi-pastural landscape planning and management, we will consider some features of modern segregationist and post-modern integration strategies for transition toward sustainable landscape development and compliant types of agriculture.

4.1.1 Features of modern strategies (IPM and IPNS)

Potential harvest

The integrated agriculture approach is aiming for highest possible production within a science based concept of 'best technical means' (De Wit, 1992; Rabbinge *et al.*, 1990). The basic frame of reference for the appreciation of agricultural production in this strategy is that of 'potential harvest'. This potential harvest is defined on basis of the 'solar energy to carbohydrate (and protein) conversion potential' of the photosynthetic system (De Wit, 1992). This energy conversion concept is a basically abstract, theoretical and generic one. Its validity has been established elegantly in microsystems, by growing micro-organisms or single crop species under laboratory conditions. However, when this model is translated to practical conditions, the potential harvest levels of any crop are not met.

Ecological consequences

In the models underlying the integrated agriculture approach, the almost 75% gap between potential and real production is explained in an interesting sequence of natural constraints. These range from production-defining factors (local radiation, temperature, and physiological features of the crop variety) via production-limiting factors (locally available water and nutrients) to production-reducing factors (incidence of pests and diseases) (Lövenstein *et al.*, 1992). This approach basically perceives nature's reality as a default situation in which environmental conditions and ecosystem interactions are competing with farmers' and societies' economic interests.

For science-advised farming practices this approach obviously enhances efforts to improve all nature's limiting conditions by technology, so as to make them meet the requirements for the highest-return crop. As a consequence of this perception, rotations are narrowed down to single crop production systems and fully controlled by off-soil conditions (hydroponics). This model brought into full practice in a radical green revolution style eventually contributes to ruining the soil, ecosystem and the rural social structures (Brown *et al.*, 1992; De Wit 1992; Hildyard 1992; Lampkin, 1990; Oldeman, 1990).

This conceptual attitude contrasts considerably with that of authors from all over the world, like Albrecht (1975), Chaboussou (1985), Dokuchayev (1892), Howard (1943), Koepf *et al.* (1996), Steiner (1924), and Vogtmann (1985). They advocate a strategy of working with nature, as a comprehensible eco-system or organism, instead of fighting against it (Baars, 1990; Bockemühl, 1984; Devall, 1985; Lovelock, 1979; Naess, 1975). From their point of view, a harvest's quality reflects the farmers' management capacities under local conditions. Also the incidence of pests and diseases is mainly an indicator of mismanagement from which farmers can learn (Kenmore, 1991). Here, a sensitive discussion starts from whether the modern, science-directed agro-technology should be regarded as mismanagement of agro-syliv-pastural ecosystems or as farming by the best technical means (Greenpeace, 1992; Hildyard, 1991; Kenmore, 1991; Vereijken, 1992).

Economic consequences

The perception of a basic opposition between economy and ecology, which is part of today's orthodox economic theory and political practice of decision making, is a conceptual one with ideological dimensions. It roots partly in the self-understanding of modern society in a competitive relationship with nature. According to this, nature is regarded as nothing but a commodity, completely at mercy of human high-handedness. Only the instrumental value of nature is being considered, and any intrinsic value is denied (Daly *et al.*, 1990; Harrison, 1994; Meadows and Randers, 1992; Schumacher, 1973; Sheldrake, 1990).

As far as the economic growth-mania is concerned, it can be realised that it is rooted in a over-emphasised attention to one of the development phases. Over-emphasis on the ego-oriented physical growth of the youth phase can easily contribute to disregard or at least depreciation of the following development phases: flowering, fructification and ripening. These phases of limited or even diminishing physical growth, offer an extensive amount of short-time and long-time 'nutritive values' to a wide range of ecosystem partners. It is interesting that in these phases integrated complexity is found in physiology and morphology. However, besides the obvious similarities, a keen awareness of the differences in the development of material and socio-cultural system should be warranted. At the moment, the actual idea of moderating physical commodity consumption is widely sensed as limiting continuing mental growth (ethical development or self-realisation). This might be seen as indicating a misidentification of physical and mental development (Bockemühl, 1984; Lievegoed, 1979; Maslow, 1968; Meadows and Randers, 1992; Schumacher, 1973). Understanding this over-emphasis and misidentification might contribute to the attitudinal shift, which is a prerequisite for the needed changes in global policy (Choudhury, 1991; Gore, 1992).

Hierarchies of interdependent networks of vital relationships between eco-partners, including humans, on the global scale of our home planet, require substantial human responsibility for our common future. Thus, in addition to the competitive economy, a concept of fraternal economy must be developed, based on the awareness to share fairly commodities world-wide. Only then, understanding and willingness can arise to adapt the humane physiological needs to match the natural limitations of global resources. This means a shift from a strategy of consuming as much as possible, financially affordable and digestible by the individual to a strategy of consuming as little as needed for the individual well-being within the social and ecological carrying capacity. The fraternal economy contrasts with the classical 'competitive' economy, originating from a social-Darwinist view of human relations (George, 1985; Van Dieren, 1995; Van Mansvelt and Verkley, 1991; Von Weiszacker *et al.*, 1997). To facilitate these urgent developments, both in ecology and economy, extensive reflections at current concepts and compliant reconsideration of ethics are required. Technical implementations can only contribute to globally sustainable resource management when they are conceived from appropriate attitudes (Blatz, 1991; Daly *et al.*, 1990; Van Mansvelt, 1988; Von Mallinckrodt, 1991).

Some prerequisites and impacts of model use

An obviously strong point of the abstract theoretical approach is that it allows, in principal, for precise analyses and subsequent mathematical modelling of a set of cause-effect relationships of the agro-ecosystems under study. It should be kept in mind, however, that modelling of the complete set is more of a theoretical than a practical possibility, even when 'all' is restricted to relevant factors. However, when we disregard the considerable time and labour input needed to fill all links of the models with legitimate data (locally, management system and crop-specific data), there is still the perception or attitude aspect of the model to be taken into consideration. This attitude consists of three interrelated parts. First, there is the presumption of the potential production level, which is a theoretical and abstract one. Strictly speaking, it is a purely ideal goal, which as such can never be reached (in practice). Second, it simultaneously implies that all features of the real world (agriculture in practice) are detrimental to the potential production. Thus, nature is perceived as a chaotic, disruptive nuisance (Gore, 1992; Sheldrake, 1990;). Third, the approach disregards the many interconnections between those sets of 'limiting' factors. Altogether agriculture is seen basically as an art to outsmart nature by means of single-factor-directed technology. The efficiency of the system is perceived as the (external) input or resource efficiency (Brinks and Van Mansvelt, 1992; Hayami and Ruttan, 1985). Generally speaking, the HEIA, IPMS and IPNS concepts are in line with this approach (FAO, 1992). In terms of sociological and anthropological trends, this radical reductionistic approach contributes to an alienation from nature (Koepf *et al.*, 1996; Van Mansvelt, 1988; Van Mansvelt and Verkley, 1991).

However, it should be added here that modelling as such is a useful tool that can create appropriate clarity on relevant issues. The point made before is that the necessary reductions, made in order to create a sufficiently simple model, can be easily forgotten when the results (predictions) of the model are communicated and used for decision making. Then problems may be caused as all aspects leaving out of the model, because of lacking data on causal relationships, are still fully in charge of reality for which the model has been designed. Apart from all kinds of linear and quantitative modelling there are also several approaches for dynamic modelling, wherein quantitative and qualitative aspects of the interacting natural and social reality studied can be linked. These models provide less hard data, but provide for a better insight in the complex interactions that actually are in charge (Struif Bontkes, 1998).

In conclusion, we can say that in spite of the obvious strong points, the outlined scientific approach require additional concepts and compliant instruments either to develop and validate the multipurpose efficiency of agro-sylvi-pastural systems. These systems are soil-bound, oriented on local carrying capacity and empirically based on the practise of minimal non-renewable resource inputs. Such sustainable agricultural systems like LEISA and Organic Agriculture (OA), are derived from the concept of co-evolution and an understandable, co-operative nature (Lampkin and

Measures, 1995; Reijntjes *et al.*, 1992; Sattler and Von Wistinghausen, 1992;). They do comply with the majority of FAO's SARD requirements (FAO, 1992; Hiemstra *et al.*, 1992; Van Elzakker *et al.*, 1992). We will discuss these options and their performances more in detail in section 3.1.2. They can be indicated as management systems, which manage agro-sylvi-pastural ecosystems as autonomous and referred to as Autonomous Ecosystem Management (AEM) (Van Mansvelt and Mulder, 1993)

Consequences for policy

In political decision making, the merging of a scientific agro-technical attitude and a 'homo economicus' approach can easily lead to proposals of segregation between agriculture and 'other' forms land use in rural areas. The modern, rational trend favours and argues for a concentration of agricultural production in the zones of high potential yield (fertile soils, favourable humidity conditions). This stand, however, tends to disregard the economical, social and cultural importance of appropriate land management in all the regions of a country.

In recent decades, the merged concepts of an economically competitive agriculture have intensified tremendously. Farmers were supposed to work towards economic optimisation, lead by technology (Anosike and Coughenour, 1990). Agriculture was aiming at the highest labour productivity, which led to intensive use of capital and commodity inputs (Hildyard, 1992; Kenmore, 1991). At the same time as this approach generated the desired increase of production output, it brought undesired side-effects of waste and labour. The approach contributed unintentionally to the degradation of soils, ecosystems, landscapes and rural societies, inside as well as outside the concerning countries. (Brown *et al.*, 1992; Greenpeace, 1992; National Research Council, 1989; Oldeman *et al.*, 1990;) The rural area's viability, the protection of environment and nature, together with an appropriate agricultural production level, require an integrated approach instead of a segregationist one. Factors such as regional infrastructure, knowledge, attitude, national price policies and (other) financial incentives should conspicuously be included in studies concerning land-use planning and evaluation (Centraal Plan Bureau, 1992; Werkgroep – De Zeeuw, 1998).

A leading and social principal in modern agricultural policy has been the perception of a basically homogeneous population of farmers. Farmers were supposed to differ only in the dimension 'advanced' (modern, rational, technically efficient) versus 'conservative' (old fashioned, romantic, inefficient). This biased point of view is misleading and should be replaced by a more accurate, differentiated one. For example, in sociological research, quality-oriented farmers have been distinguished from quantity-oriented farmers, and capital- and/or technology-intensive farming from labour- and/or knowledge-intensive farming (Van der Ploeg and Ettema, 1990). Here, not only the local differences (natural and infrastructural) but also those in style of farm management (mental and psychological) are taken into account (De Bruin *et al.*, 1992; Roep *et al.*, 1991; Van der Ploeg and Roep, 1990). In all farmer groups more and less financial successive farmers were found. Labour diversification on farm and job diversification in the region, might be post-modern solutions by which farmers can enhance ecological and social sustainability as well as economic viability. Empowerment of regional specificity and increased access to appropriate education in the region are key issues in the development of diversified management of agriculture and related crafts (Anosike and Coughenour, 1990; Evans and Ngau, 1991; FAO, 1991). Shifting away from the idea of generic uniformity to that of structural diversity will most probably favour a more flexible, sustainable policy-making and will finally have a clearly visible effect on the region's landscape.

To use of the checklist, as presented in Chapter 3, by users and for uses, as referred to in Chapter 5, means that it should always be possible to find an optimal balance. This balance should be between the generic objectives, which are valid throughout the EU or even wider, and the local conditions, which have to be considered. The prevention of soil erosion is a generic aspect, with appropriate soil covers and management as a generic tool to reach it. However, the appropriate kind(s) of vegetation(s) are to be decided by the local landscape management experts. Their expertise should not only be disciplinary but include the capacity to discuss the various options in the context of a management that is sustainable and supports the ongoing development of the regional identity in a national and international context.

4.1.2 Basic concepts of organic types of agriculture

Autonomous Ecosystem Management

Like all types of agriculture, organic types of agriculture are implementations of a basic concept, model or mental map. In organic types of agriculture, the conceptual common denominator can be defined as autonomous ecosystem management (AEM) (Van Mansvelt and Mulder, 1993). This includes such notions as optimising the primary production efficiency of agro-ecosystems, in compliance with the local soil and climate conditions (carrying capacity) and the social needs of the region. In view of the management requirements, this strategy of post-modern agriculture demands an attitude in favour of an exchange of eco-intelligence for non-renewable resources. Agriculture, in this concept, is more a policy for land-use in general, including agro-, sylvi- and aqua-culture in mixed or integrated agro-ecosystems. Pest prevention and well-balanced mineral flows and sustainable resource management result in low external inputs (chemical fertilisers and non-renewable energy). This basic conceptual framework of agricultural systems management can be traced back in many countries' agronomic literature. Together with many of the aspects indicated in following sections, they have been explained by researchers such as Albrecht (1975), Boehnke (1992), Dokuchayev (1892), Draghetti (1991), Howard (1943), Kenmore (1991) and Steiner (1924).

Cultivating the multipurpose efficiency

To optimise the production of agro-ecosystems in the framework of AEM, it is important to cultivate the multipurpose characteristics of the relevant kingdoms of nature - the eco-partners of the system. Over-stressing any single production aspect of any subsystem might easily lead to the deterioration of the balanced efficiency of the whole system. Orchestrating the benefits of diversity (Van Elzakker *et al.*, 1992) or optimising the synergy of soil-crop-animal interactions, is therefore the foremost challenge of organic types of agriculture (Van Mansvelt and Verkley, 1991). This strategy includes using animals to upgrade the non-foods produced with the foods, and to provide manure for the soils to produce food, feed and fibre (Koepf *et al.*, 1996; Lampkin, 1990; Sattler and Von Wistinghausen, 1992).

Within this framework, some aspects of the multipurpose efficiency of eco-subsystem can be listed as follows:

1. Soils (mineral-clay-humus complexes and sandy, loamy or peaty) merge various purposes:
   - Source and store or buffer of nutrients (i.e. organic and inorganic nutrients), and water for crops (micro, meso and macro fauna);
   - Bio-tope for general waste-feeders (for C/N balance, nutrient transformation and waste recycling). Here it becomes obvious that the idea of waste/garbage as non-useful matter is largely a lack of consistent thinking in resources' life-cycles;
   - Basis of the agricultural production (sustainable soil-fertility, sustainable land-use, land regeneration and/or improvement).

2. Crops (mainly floriferous plants) merge various purposes:
   - Human food and fibre producers (carbohydrates, oils and proteins, fuel, shelter, building and clothing materials);
   - Animal feed producers (energy and protein, shelter, bedding);
   - Soil organic matter and structure producers (leguminous and bacterial N-fixation, edaphon feed/energy input, soluble minerals (nitrogen) trapping, soil structure and mineral mobilisation). It is important to be aware that, in this case, the N input in the system goes intrinsically together with an input of carbohydrates, which provide structure (root action) and an input of energy for the edaphon (phases in root decay) to the soil;
   - Water harvesters (water uptake through root canals, dew or air moisture interception by crops or treetops, and reduction of evaporation);
   - Climate regulators (mico-, meso- and macro-climate; wind shielding, shadow casting, temperature moderating and rain catching.
3. Husbandry (mainly vertebrates, but also fowl, fish and bees) merges various purposes:
   - Human food and fibre producers (milk, meat, honey, hair, wool, hides and wax);

- Roughage feeders (non-food to food transformation, waste feeders instead of competitors for human food). Ideas on dairy feed regimes have recently been adapted away from over-emphasising the importance of protein to re-appreciation of energy (carbohydrates). In this way the gradual degradation of the food has become important as well as the sheer nutrient uptake;
- Manure production (N recycling and redistribution and stabilising of soil aggregates);
- Intelligent traction (low external energy input and efficiency dependent on infrastructure);
- Pollination.

4. Climate (sunshine, rain, temperature and wind) merges various purposes:
   - Provides time-scale and trigger for the regulation of development phases (long-time, seasonal, and diurnal cycles);
   - Source of external (solar) energy input for the primary (plant) production.

Below, the main before mentioned keywords will be explained in some more detail:

Aspects of soil management

In agreement with farming tradition in many regions, organic types of agriculture regard soil improvement, by means of well-balanced land-stewardship, as a key issue of its professional ethics. Regeneration and conservation of the soil's fertility are basic requirements as well as a challenge for the farmers' craftsmanship. In this framework, providing the successor with a better soil is the ultimate goal. Appropriate liming and rock-dust applications are accepted as medication of soils in need of special care (for example nutrient deficiencies). Considerations of non-renewable energy and other resource depletions are therefore critical, stressing the need for minimal input strategies. On site nutrient mining by deep rooting crops, improvement of nutrient availability with mycorrhizas and optimal nutrient recycling are components of the multipurpose approach. Within the agro-ecosystem, specific crop rotation and manure strategies for different soil types, structures and exposures are part of the craftsmanship.

Aspects of crop rotation

Within the framework of the multipurpose efficiency of crop production, crop rotation is instrumental to orchestrating the complementary characteristics of various crops in their mutual and plant-soil relationships. Inter-cropping, after-cropping, alley-cropping and mixed-cropping are examples of spatial and temporary alternations, always chosen according to crop- and soil-specific cycles, needs and gifts. Sophisticated crop rotations do also appear instrumental in biological pest prevention (fungi, insects and weeds). Additional multipurpose aspects of crops, to be considered when designing rotations, are for example:
- Legumes as N-fixers and protein food/feed producers;
- Leguminous ley-grasslands as N-fixers, feed producers and weed oppressors;

- Corns and grains as producers of (staple) food and feed, animal housing and a N-absorbent material for all kinds of manure (balancing the C/N ratio back to soil-appropriate levels);
- Vegetables (food diversification) as vitamin and mineral sources for humans and husbandry;
- Roots as (staple) food, feed and silage;
- Fruits (fresh and staple, also feeding pest predators);
- Herbs (teas, spices, medicine);
- Flowers (feeding beneficial insects and producing colour in the landscape);
- Seeds (propagation, elevation, oils);
- Hedges and Woodlands (feed, shelter, housing, burning, landscaping, soil stabilisation, humidity regulation).

Aspects of mixed husbandry

Within the framework of multipurpose efficiency of animal husbandry, cultivation of the animal-specific characteristics is instrumental. As before, implementation always depends on the actual farm situation, soil and climate conditions, and inter-species interactions of local or adapted breeds with non-husbandry species. The multipurpose approach of animals focuses on the manure production and the roughage and waste-to-food conversion capacity of cattle and other animals. These features are important keys to avoid competition on human food in a world of limited resources (Meadows, 1992; Meadows and Randers, 1992).

Cultivation of cycles and developmental phases

To optimise or 'orchestrate' the multipurpose potential of the before mentioned eco-subsystems, the whole scale of different qualities of developmental phases, seasons and other cycles must be taken into account. The features to be considered include the following:

1. Specific properties of the seedling, growing, flowering, ripening, decaying and dormant phases in crops:
    - Selective cropping for food, feed, fuel and timber in forestry (including crops for green manure which are harvested before flowering and special after-sowing of crops for nitrate trapping);
    - Sustainable seed production (on-farm and/or local).
2. Specific properties of the young, adult and mature phases in animals:
    - Meat versus reproduction;
    - Dairy versus traction;
    - Health and long-life breeding versus stressful top-seed growth.
3. The alternating consumption of plant-food in the growing season and animal-food in the 'hunger' season;
4. Management anticipating on bio-meteorological cycles (such as sunspots and locust plagues, lunar cycles and cassava planting).

Farm management: privatisation and associative co-operation

Reflecting on the listed options and considering the requirements for professional craftsmanship, it will become clear that agriculture in general and organic agriculture in particular demands great knowledge and ability. These days, most farms in the western parts of Europe, although indicated as family farms, are run by one person and clearly show that person's preferences in scale and emphasis on vegetable or arable crops, small or large livestock, fruits or whatever. This situation is often perceived as a deadlock of either professional farming, albeit highly specialised, or within mixed farming, although this is relatively unprofessional. Both would structurally exclude the options of multipurpose efficiency as presented. To overcome this conceptual and habitual deadlock, the establishment of associative forms of farmer's co-operation between farmers might be seen as an interesting tool, which would be instrumental in helping the farmers to profit from the benefits of diversity. On the one hand, it can give the farmers a larger say in the agro-business complex: it will make them a stronger and more independent partner. On the other hand, it could serve the appropriate task division within the mixed, multipurpose types of agriculture discussed.

In our opinion, associative community farming and co-operation of farmers in general, could considerably increase the viability of rural life. By providing a critical mass for renewed rural community building, it sets an end to the isolation of the farmers, the spread of which is a key issue in rural degradation. Various forms of co-operation within the food chain, linking farmers to consumers, are necessary to overcome the huge gap of alienation between farmers and consumers (De Bruin *et al.*, 1992; Gengenbach and Limbacher, 1991; Groh and MacFadden, 1990; Hiemstra *et al.*, 1992; Klett, 1990).

4.1.3   Concluding remarks

From the considerations mentioned above, it can be clear that the concept of organic agriculture and the attitude in which they root do fully comply with the requirements for a sustainable land-use, merging the care of a healthy development of the land and society. Healthy food and fibres, renewable energy and landscape satisfy human's need for nutrition, recreation and aesthetic experiences and are organically inter-linked with products of multifunctional land-use systems. The basic concepts of organic agriculture are addressed as aspects of one major target. However, the standards of organic agriculture refer only to the concepts of organic agriculture as far as they are directly related to the sustainable production of food and fibre. The issues of nature and landscape production are indicated, but not yet fully reflected within those standards. For recent data on the resource use in organic agriculture see Isart and Llerena (1997).

4.2   EMPIRICAL DATA COLLECTED FROM LITERATURE

Some years ago, Van Mansvelt and Mulder (1993) published an overview of the performances of organic agriculture. At that time, mainly disciplinary studies were available, looking into more or less depth of particular aspects of the organic and non-organic (conventional) farming systems. Based on several publications quoted by Van Mansvelt and Mulder, performances of organic agriculture are found on the following issues:

1. Nutrient leaching into the environment was found to be definitely lower in organic agriculture than in conventional agriculture and largely complying with the requirements of the EU for drinking water. Since the publication of Van Mansvelt and Mulder (1993), several other authors published more recent results underpinning the same conclusion (CLM, 1997).
2. The loading of pesticides into soil, water and air was presumed to be not relevant to compare at all as the use of pesticides is excluded by the standards of organic agriculture. However, in the case of copper (oxychloride) against leaf-fungi it is different. Similarly, plant or microbe-based biological pesticides are not completely excluded either. So, although the use of pesticides is minimal and the breakdown of most active substances is fast and complete, the issue is not to be neglected (CLM, 1997). Derivatives leaving from organic agriculture are well below the levels of conventional farm practices in many agricultural branches like grapes, horticulture and fruits.
3. Yield volumes are lower in organic agriculture than the maximum yields of conventional agriculture, especially in developed countries. However, the latter yields often go together with considerable losses in nutrients and environmental quality and cause national food surpluses, which are exported with subsidies. Such export subsidies can cause disruption of local markets and rural areas elsewhere. Especially, in case of intensive animal production the competition for food and thus of available surfaces for food production between husbandry and human beings is considerable. Moreover, intensive off-soil husbandry affects landscape and nature negatively and contributes with high animal protein consumption to human health problems. Studies about national food security and sufficiency show that, based on the WHO recommended daily nutrient intake, organic agriculture can well feed the populations of developed countries (Van Mansvelt and Mulder, 1993).
4. Species and habitat diversity were well covered in the quoted studies in Van Mansvelt and Mulder (1993) and indicating very strongly that organic agriculture is favourable for fauna and flora on micro, meso and macro level, up to birds and mammals in soil, water, and many vegetation types. More recent studies support and elaborate this feature of organic agriculture, which is very important for sustainable landscape management (Smeding, 1995; Vereijken, 1996a, 1996b).
5. Regarding the economics of organic farming, the studies referred to by Van Mansvelt and Mulder (1993) show that the differences between economic

148

performance of organic farmers and non-organic farmers are much larger within both groups than between them. More and more evidence appears that organic farming is as least as profitable as non-organic.The economic perspectives for organic agriculture will arise if financial support is given to the conversion of conventional to organic agriculture in the EU and organic agriculture's performance on environmental and ecological issues is rewarded. Efforts from the European organic agricultural organisations to have a lower VAT level agreed for their products is an example, which will support the economic perspective of organic farming. A lower VAT level will reward the lower level of social and environmental costs caused by organic agriculture.

6. The society aspects of organic agriculture focus on the efforts of organic agriculture to reconnect agriculture with consumers, who became very alienated from agriculture on which they rely for the quality and quantity of their food. It becomes also more and more accepted to discuss farmers' labour diversification of on- and off-farm labour. Moreover, it becomes clear that there is not a one-and-only winning strategy for farming. There are various possibilities to be a successful farmer (Roep et al., 1990).

The general conclusions from the publication of Van Mansvelt and Mulder (1993) mentioned above are still realistic and true. Organic agriculture is a fully feasible option for sustainable rural development as required in all policy papers regarding food, fibre, energy and landscape production for the next century. It is fair enough to say that organic agriculture is not a panacea and still needs considerable effort to stay ahead. Besides that, options for organic agriculture need increased support from policy and (interdisciplinary) research in order to realise its promising potentials. In that respect, it is like all promising perspectives, viz. they need full, serious, compatible and well-organised support in order to get implemented.

From a landscape point of view, on which the concerted action was based on, it strikes that social, aesthetic and landscape physiognomic studies about landscape development and organic farms are largely missing. The reason given for this lack is that organic farms are so small in number that they do not fit in any representative sampling. Such an argument will exclude any research of the various disciplines on new agricultural strategies and thus it can be considered as an interesting phenomenon of research policy with little scientific weight. Realising the number of disciplines finding organic agriculture to be feasible and favourable for the requirements of sustainable management, it can be presumed that organic agriculture indeed has a key to merge targets in a synergetic and compatible way. Interdisciplinary studies, like this one, supports and specifies such preliminary conclusions. Obviously, new studies should improve the methods applied in the concerted action. This means a shift from quick scans to long term observations, resulting in useful proposals for organic and non-organic farmers to improve their management.

4.2.1 Comparison of farming systems in the concerted action

During the EU concerted action, "The landscape and nature production capacity of organic/sustainable agriculture", interdisciplinary research teams of various countries made quick scans comparing the landscape production of organic farms with the landscape production of neighbouring non-organic farms. As these research teams looked at all major scientific aspects, in a comprehensive way, their results contribute to the indications of the above referred disciplinary studies and support the conclusions about the compatible synergy, on organic farms, between the targets set by the values of the various disciplines.

Hendriks and Stroeken (1992) studied the production of landscape on four bio-dynamic farms in three countries. They found that the four bio-dynamic farms showed more diversity and a better coherence than the non-organic neighbours. They also found that that the bio-dynamic farms had a better performance on all studied realms, viz.: ecological diversity, landscape diversity, product diversity, labour diversity and sensorial diversity. Thus the bio-dynamic farms indicated a synergy between multiple targets.

Rossi *et al.* (1996, 1997) made a quick scan of the landscape production on two organic farms in Tuscany and compared these results with the landscape production of non-organic (conventional) farms in the surroundings. Their main objective was to assess the feasibility of a third version of the checklist of the EU concerted action. Their critical and constructive comments on the checklist were discussed during the next meetings and were partly integrated in the next versions of the checklist. However, taken the pilot character of the research into account, the following tendencies can be concluded from their results about the scores of more than 140 parameters presented on the list. They used a five points validation scale (++, +, +/-, -, --). The organic farms scored on all parameters ++ or +, while the conventionally farmed surroundings scored +/- or − respectively. So, obviously the organic farms contributed better to the sustainable management of the landscape. Looking at the scores of the subsequent columns, the clearest discrepancies were found for the columns 1 (environment), 2 (ecology) and 6 (physiognomy and cultural geography) and showed that all three organic farms had better scores than the conventional farmed surroundings. One of the organic farms had also a better score on the parameters of column 3 (economics) and 5 (psychology), while the other two organic farms had a better score on column 4 (sociology). None of the produced landscape on organic farms had a lower score on any of the columns compared with the landscape produced by conventional farms in the surroundings. One of the organic farms, which was situated in the less sustainable surroundings had a lower score on column 5 (psychology) than the other two organic farms, which were situated in more diverse and coherent surroundings. Referring to the question of synergy between the columns of the checklist, a guarded conclusion is that no clear incompatibilities have been found. This may for the time being, be taken as a indication that, at least within the applied concept of organic agriculture, a synergy between the targets of the six columns may exist.

Pauwels *et al.* (1996) compared two organic farms in northern Belgium with the neighbouring non-organic farms and assessed the feasibility of the third version of the checklist of the EU concerted action. They gave constructive comments, which were also discussed in the next meetings and partly integrated in the next versions of the checklist. They concluded that, although the effects of the organic farms on the landscape were small, because of their small number in Belgium, their landscape performance as such was better than that of the non-organic farms, because of the mixed character and environmental friendly management of organic farms. They also noticed that landscape standards are not yet included in the standards for organic agriculture, which makes the efforts for landscape production more dependent from the farmers' personal motivation.

Kuiper (1997) emphasises that one individual farmer or farm can only make a small contribution to the landscape. So, few organic farms in a landscape, even if they perform better according to the targets and criteria on the checklist than conventional farmers, will still have little effect on the landscape as a whole. Moreover, if the number of organic farms would increase in a landscape and they all would follow a similar type of farm design, then the landscape will be more diverse at the level of species, but perhaps remains monotonous at the level of habitat-distribution. Such a consequence makes a strong point to support the idea of an integrated approach to regional land-use, starting on watershed or community level. Co-operative management of rural landscapes as agro-sylvi-pastural land-use units in a context of sustainable development, is one of the challenging perspectives indicated in Kuiper (1997).

Hendriks *et al.* (1997) presented a comparison between two organic and two non-organic farms in West-Friesland, the clay-soiled 'cabbage' area of The Netherlands. The non-organic farms were environmentally aware (MBT) farms. Hendriks *et al.* (1997) agree with Pauwels *et al.* (1996) that the standards for organic agriculture derived from IFOAM (1996), EU (2091/92) and SKAL (1997), do not yet warrant anything else than a sustainable level of emissions, a certain state of animal welfare and soil fertility. This means that for landscape, crop rotation and non-use of herbicides are the major sources of diversity and appear mainly on the species level. Plantation of woody elements and additional habitats are not a requirement for the label, but they comply with the concept of organic agriculture and eager farmers will be inspired to put that concept into practice. However, farmers with other priorities may stick to the minimal requirements. Non-organic farmers that are eager about ecology and landscape values may reach many of the targets on the checklist and do better than some organic farmers. The area studied by Hendriks *et al.* (1997) is specialised in horticulture, flower-bulb production and mixed farming. A preferred type of organic farming is absent in this area.

In another area of the Netherlands, called Waterland, between Amsterdam and West-Friesland, Hendirks *et al.* (1998) compared organic and non-organic farms in a peat area, which is mainly used for dairy farming. The comparison has been executed in the same way as the farms in West Friesland mentioned above. One of the special features of this landscape research is that the visual appearance of landscape in each of the four main seasons of the year has been assessed, arguing that landscape value is not a snap-shot issue, but a source for impressions affecting daily life all year round. The cyclical changes of landscape appearances and the temporal coherence of the successive seasons, belongs to the appreciated features of landscapes (Coeterier, 1996). Besides the temporal coherence, organic farms also showed a strong vertical coherence and a higher level of dependency on local soil and water conditions, because of their low-external-input policy.

4.2.2   Concluding remarks

From the above presented studies about the actual landscape performances of farmers, it is clear that the landscape performance of any farm must be seen in its surrounding landscape and that there are considerable differences in the quality of landscape production between organic and non-organic farmers. There is also considerable difference in the quality of landscape production among organic farmers and among non-organic farmers. At the same time quite clear indications exist that organic agriculture has considerable advantages related to the landscape quality, because organic agriculture is soil-based (on-soil farming), has a low-external-input approach, goes for mixed farming, has wide crop rotations wherever possible, and largely refrains from pesticides. Moreover, organic agriculture tends to attract farmers, who are aware of the contribution of their farm to the landscape quality. Such farmers are willing to take compliant action. Addition of some feasible landscape standards to the standards of organic agricultural production would be quite feasible and acceptable to show the effects of cross-compliant farming on the EU schemes of farmer income support, which are necessary to maintain and develop the European rural landscape in a sustainable way.

CHAPTER 5   USES AND USERS

At the beginning of the concerted action AIR3-CT93-1210 "The Landscape and Nature production Capacity of Organic/Sustainable Types of Agriculture" the main target has been to provide policy with guidelines for sustainable landscape management (see also the objectives in section 1.3). This original target of the concerted action was focused on the international policy of the EU institutions and the European Working Group of the International Federation of Organic Agriculture Movements (IFOAM). They were supposed to be the optional users and to use the guidelines as a setting for standards and payment schemes. During the concerted action and especially during the last years when the participants of the concerted action co-operated with local experts in the countries visited, the envisaged uses and users have been discussed over and again. In face of the wide variety of environmental, ecological, economic, social and cultural conditions in the countries and regions visited, the idea of phrasing parameterised standards, which are generally applicable for regional landscape management and add up significantly to the existing EU or IFOAM standards for organic agriculture, faded away. However, the importance of explicating the contexts leading to whatever criteria and parameters set has been realised again. It became clear that to meet targets criteria and parameters are necessary to assess how good the targets can be reached. It also became clear that to find and to meet parameters instead of criteria and criteria instead of targets, inevitably leads to counterproductive biases. This could be illustrated by all single target payment schemes, in which targets are set by parameters. For instance sheep-per-head payments, hedge-length payments, commodity volume payments, bird-nests per surface area payments. Therefore the focus of this study has been shifted towards the establishment of a checklist for sustainable landscape management instead of the development of standards for sustainable landscape management. An explanation of the targets and the suggestions for and development of criteria and parameters seems to be much more appropriate for an EU-wide use. Especially as one of the major overall targets is to contribute to the care taking of and further development of the regions' identities in their characteristic diversities in Europe. Realising that local commitment is a paramount for a policy considering regions' identity, it has been decided to present the checklist as in chapter 3, which could be useful for all stakeholders involved in landscape management wherever and on whatever scale.

It has been kept in mind that even when very precise standards are developed at a general level and to be implemented into practice, implementation will be open to such interpretations that may considerably differ from place to place and from user group to user group. (See also Consequences for policy in 4.1.1).

5.1 OVERVIEW OF POSSIBLE USES AND USERS

In the last year of the concerted action, during the meeting in Crete, in which the regional and international experts participated, an extra workshop about uses and users of the checklist was organised. During that workshop, a list has been created with possible users and uses of the checklist. The following table has been derived, which indicates in keywords the possible uses each of the possible users can make of the checklist. Obviously, the list is only meant as indicative and should in no way be taken as limiting either in uses or the users. The potential uses and user-groups can be at various scales and levels, viz.: international, national, regional and local.

Table 5.1: Examples of possible uses and user-groups of the checklist.

Uses for Users>>	Politicians and Administration	Advice and Extension	Research	Education	Farmers	NGO's/interest groups
Checklist	Payment control of sustainability schemes	Relevant (rural) landscape issues	Ongoing updating and refining	Teaching and Examination	Regional codes of good practice, farm validation	Checking institutional planning
Design framework	Identify and validate land use and agriculture	Consistent programming and planning	Interdisciplinary methodology design	Curriculum design	Farm design, development and management	Choosing priorities and synergies in planning
Strategic planning	Policy targets for sustainable land use	Translation to farming practices	Unifying concept's development	Strategic planning of Education	Farmers' association's strategic planning	Integrating own objectives with other groups'
Communication	Discussions among interest groups / stakeholders	Awareness raising in planning and management	Interdisciplinary research projects	Interconnection of teaching topics	Communication with other land-users / interest groups	Communication with farmers and other interest groups
Financing	Grants & Cross Compliant income support	Farmers financing services	Efficient & appropriate payments	Educational financing	Income diversification	Lobbying for the financing of sustainable land-use.

5.2 INDICATIVE LINKS WITH FUNDING

During the discussions about possible uses and users, also possibilities for funding landscape management along the lines of the checklist have been looked for. There are two different strategies to fund landscape management, viz.: target funding and procedure funding. Here, target funding is to focus on targets at a sufficiently general level of integration to warrant its appropriate efficiency.

1. Target funding:
- Funding of agro-landscape production, which develops and/or maintains features of the specific local or regional qualities that are characteristic for the landscape's identity.
- Funding of farmers through income support, adding up the fall in income from a reduction of food-sales caused by the world-wide open market and the exteriorisation of costs of environmental protection and landscape maintenance.

2. Procedure funding:
- Funding of NGO groups of local farmers, land–users and all other relevant groups of stakeholders involved in and committed to sustainable landscape and land-use planning. This strategy raises local awareness of the real multifunctionality of the landscape and enhances participation in this issue. Thus funding of NGO groups contributes in two ways to the empowerment of the rural population.
- Funding of pilot conversion projects for sustainable land-use and landscape management. This is important to show how feasible strategies for these targets can be implemented in a way that meets the targets and fits to the local conditions of the criteria and parameters of the checklist, in that particular region and at that particular scale.

For all such funding programs, the criteria and parameters of the checklist can be used in general and specified with the knowledge of a local expert panel, to warrant appropriate application. The local panel consists of disciplinary experts with supra-regional expertise, which are peremptory to warrant that the figures of the region are considered appropriately in its context on the next scale in the hierarchy of systems. To decide on the relevance of stakeholders to be invited in planning and decision-making on the landscape's future, it seems crucial to look for a fair balance between give and take or between rights and duties, in a perspective of proportionality. Thereby, an acceptable balance between individual liberty and common fraternity or between freedom and responsibility should be sought after in an atmosphere of equal rights. Being definitely aware of the arbitrariness of the criteria for decision-making, it seems that making criteria explicit and discussing them openly with all stakeholders involved is a promising option to arrive at a sufficient degree of transparency and acceptability of the decisions and to warrant their appropriate application in practice (Volker, 1997; Bosshard and Eichenberger, 1998).

6   REFERENCES

Achterhuis, H.J., 1992. *Technologie en folosofie: beeld en werkelijkheid.* Enschede: University of Twente, 24p.

Agnew, A.D.Q., Collins, S.L. and Van der Maarel, E., 1993. *Mechanisms and processes in vegetation dynamics.* Uppsala: Opulus, 134 p.

Albrecht, W.A., 1975. *The Albrecht papers.* Raytown, U.S.A.: Acres, 401 pp.

Alfoldi,T., Spiess, E., Niggli, U. and Besson, J.M., 1995. *Energy input and output for winterwheat in bio-dynamic, bio-organic and conventional production systems.* Ashford: Ashford Wye college, 680p.

Altieri, M. (ed), 1992. Special issue: Sustainable agriculture. *Agriculture, Ecosystems & Environment*, 39 (1-2).

Altieri, M. A., 1995. *Agroecology: The science of sustainable agriculture.* London, Westview Press, IT publicdations, 433p.

Altieri, M., 1989. Agroecology: A new research and development paradigm for world agriculture. *Agriculture, Ecosystems & Environment*, 27, pp. 1-24.

Anonymus, 1991. *Using old breeds of domestic animals in landscape management.* Proceedings. Hessen: Wetzlar, Naturschutz Zentrum, 34pp.

Anosike, N. and Coughenour, C.M., 1990. The socio-economic basis of farm enterprise diversification decisions. *Rural Sociology*, 55 (1), 1-24.

Ansaloni, F. and De Roest, K., 1997. Use of resources and development and organic livestock farming in Italy: first results of an ongoing study. In: Isart, J. and Llerena, J.J. (eds). *Resource use in organic farming.* Proceedings of the third ENOF workshop, Ancona, 5-6 June, 1997. Barcelona: The European Network for Scientific Research Coordination in Organic Farming (ENOF).

Audiot, A. and Flamant, J.C., 1992. Traditional animal breeds: economic asset of the South-West part of France. *Bulletin d' Information sur les Resources Genetiques Animales.* Rome: FAO/PNUE, 10, pp. 33-40.

Audiot, A., 1995. *Using yesterdays' breeds for future breeding.* Paris: INRA, 230pp.

Baars, T. and Bloksma, J., 1995. *Dynamisch onderzoek voor biologische landbouw.* Driebergen: Louis Bolk Instituut, 27p.

Baars, T. and Buitink, I., 1995. *Enkele praktische aspecten van ziektepreventie in de biologische veehouderij: gezondheid ondersteunen, natuurlijk gedrag bevorderen.* Driebergen: Louis Bolk Instituut, 74p.

Baars, T., 1990. Het bos-ecosysteem als beeld voor het bedrijfsorganisme in de biologisch-dynamische landbouw. Driebergen: Louis Bolk Institute, 32p.

Baggerman, T. and Hack, M.D., 1992. *Consumentenonderzoek naar biologische produkten: Hoe het marktaandeel vergroot kan worden.* Mededelingen 463. The Hague: Agricultural Economics Research Institute (LEI-DLO) and Consumer Research Institute (SWOKA).

Bakker, Th. M., 1985. *Eten van eigen bodem, een modelstudie.* The Hague: LEI-DLO, 255 p.

Baldock, D. and Beaufoy, G., 1993. *Nature conservation and new directions in the EC common agricultural policy.* Report for the Ministry of Agriculture, Nature Management and Fisheries, the Netherlands. Arnhem and London: Institute for European Environmental Policy, 224 p.

Baldock, D. and Mitchell, K., 1995 (draft version). *Cross compliance within the common agricultural policy: A review of options for landscape and nature conservation.* A discussion document for the Netherlands' Ministry of Agriculture, Nature Management and Fisheries and the UK Department of the Environment. London: Institute for European Environmental Policy, 81p.

Ballieux, P. and Scharpe, A., 1994. *Organic agriculture*. Brussels: Publication Office of the European Community, 41 p.

Barret *et al.* 1990. Cited in Barret, 1992.

Barret, G.W., 1992. In: Olson, R.K. (ed.) *Integrating sustainable agriculture, ecology and environmental policy*. The Haworth Press.

Barrow, C.J., 1991. *Land Degradation*. Cambridge: Cambrige University Press.

Batteman, D.I., 1993. Fiancial and economic issues in organic farming: A case study in pluriactivity. *Journal of the University of Wales*, 73, pp. 4-31.

Baumol, W.J. and Oates, W.E., 1988. *The theory of environmental policy*. Cambridge: Cambridge University Press.

Beismann, M., 1997. Landscaping on a farm in northern Germany, a case study of conceptual and social fundaments for the development of an ecologically sound agro-landscape. *Agriculture, Ecosystems & Environment*, 63 (2/3), pp. 173-185.

Bennet, G., Von Weizsaecker, E.U. and Baldock, D., 1990. *The international market and environmental policy in the Federal Republic of Germany and the Netherlands*. 's-Gravenhage: VROM, 123 pp.

Birdlife International, 1997. *An Agenda for Action: Reform of the CAP*. Brussels: Birdlife International.

Blatz, V. (ed) 1991. *Ethics and Agriculture*. Moscow: University of Idaho Press, 138p.

Bockemühl, J., 1984. *In Partnership with Nature*. Wyomin: Biodynamic Literature, 115p.

Bockemuhl, J., 1992. *Erwachsen an der Landschaft*. Dornach: Naturwissenschaftliche Sektion, 320p.

Boehncke, E., 1992. *Basic principles of organic animal husbandry*. Paper presented at the conference on New strategies for sutainable rural development. Godollo University of Agriculture Sciences, Godollo, 22-25 March, 1992.

Boehncke, E., 1997. Preventive strategies as a health resources for organic farming. In: Isart, J. and Llerena, J.J. (eds). *Resource use in organic farming*. Proceedings of the third ENOF workshop, Ancona, 5-6 June, 1997. Barcelona: The European Network for Scientific Research Coordination in Organic Farming (ENOF).

Boer, K., 1993. *Ecologisch groenbeheer in de praktijk*. Arnhem: IPC Groene Ruimte.

Boisdon, I. And L'Homme, G., 1997. Global mineral balance on diversified organic farming systems. Prototypes. In: Isart, J. and Llerena J.J. (eds). *Resource use in organic farming*. Proceedings of the third ENOF workshop, Ancona, 5-6 June, 1997. Barcelona: The European Network for Scientific Research Coordination in Organic Farming (ENOF).

Bojo, J., Maler, K.G. and Unemo, L., 1990. *Environment and development: An economic approach*. Serie Economy & Environment. Dordrecht: Kluwer Academic Publishers Group.

Booij, C.J.H. and Noorlander, J., 1992. Farming systems and insect predators. *Agriculture, Ecosystems & Environment*, 40, pp 125-135.

Bosshard, A. (in preparation). *Assessment of sustainability: A conceptual framework*. Tann: Office for Ecology and Agriculture.

Bosshard, A. 1997. What does objectivity mean for analysis, valuation and implementation in agricultural planning? A practical and epistemological approach to the search for sustainability in 'agri-culture'. *Agriculture, Ecosystems & Environment*, 63 (2/3), pp133-145.

Bosshard, A., Eichenberger, M., Eichenberger, R., 1997. *Nachhaltige Landnutzung in der Schweiz: Konzeptionelle und Inhaltliche Grundlagen fur ihre Bewertung, Umsetzung und Evaluation.*

Bouma, J., Verhagen, J., Brouwer, J. and Powell, J.M., 1997. Using systems approaches for targeting site-specific management on filed level. In: Kropff, M.J., Teng, P.S., Aggarwal, P.K., Bouma, J., Bouman, B.A.M., Jones, J.W. and Van Laar H.N. (eds.) *Application of systems Approaches at the Field Level.* Proceedings of the Second International symposium on Systems Approaches for Agricultural Development, held at IRRI, Los Banos, Philippines, 6-8 December, 1995. vol.2. Dordrecht, Boston, London: Kluwer Academic Publishers.

Brinks, G. and Van Mansvelt, J.D., 1992. *The influence of market and government policy on resource use and environment in agriculture.* Wageningen: Wageningen Agricultural University. Paper for UNCTAD expert consultation.109p.

Brown, L.R., Flavin, C., Postel, S, and Starke, L. 1992. *The State of the World.* Washington: Worldwatch Institute.

Brundtland, G.H., 1987. *Our common future.* World Commission on Environment and Development. Oxford: University Press.

Brundtland, G.H., 1992. *Statement at the opening of the conference on Environment and Development,* Rio de Janeiro, 3 June.

Bujaki, G., Guzli, P. and McKinlay R.G., 1995. Comparison of energy output/input of conventional and organic agriculture in Scotland and Hungary. Farnham, BCPC Symposium Proceedings, no 63, pp 179-182.

Buys, J.C., 1995. *Naar een natuurmeetlat voor landbouwbedrijven.* Utrecht: Centre for Agriculture and Environment.

CEC-DG XI (Commission of the European Communities, Directorate General XI-Environment, Nuclear Safety and civil Protection), 1993. *Towards Sustainability: a European Community programme of policy and action in relation to the environment and sustainable development.* Luxemburg: CEC-DG XI.

Centraal Plan Bureau (CPB), 1992. *Scanning the future. A long term scenario study of the world economy, 1992-2015.* The Hague: SDU, 328p.

Chaboussou, F. 1985. *Sante des Cultures: une Revolution Agronomique.* Paris: La Maison Rustique, 126p.

Chamberlain, D., R. Fuller and D. Brooks, 1996. The Effects of Organic Farming on Birds. *Elm Farm Research Centre Bulletin,* January 1996, 21, pp: 5-9. Newbury, RG20 0HR, UK: Elm Farm Research Centre.

Choudhury, K. 1991. Intervention at the FAO-SARD conference, Den Bosch (NL).

CLM (Centrum voor Landbouw en Milieu), 1997. *Milieuprestaties van Eko-Landbouw.* Utrecht: Centre for Agriculture and Environment.

Coeterier, J.F., 1996. Dominant attributes in the perception and evaluation of the Dutch landscape. *Landscape and Urban Planning,* 34, pp: 27-44.

Colmenares, R. and De Miguel, J.M., 1997. Landscape perception and grassland management in Central Spain: From traditional livestock raisers to modern organic farming. In: Isart, J. and Llerena J.J. (eds). *Resource use in organic farming.* Proceedings of the third ENOF workshop, Ancona, 5-6 June, 1997. Barcelona: The European Network for Scientific Research Coordination in Organic Farming (ENOF).

Colquhoun, M. 1997. An exploration into the use of Goethean science as a methodology for landscape assessment: the Pishwanton project. *Agriculture, Ecosystems & Environment,* 63 (2/3), pp. 145-159.

Constanza, R. (ed), 1991. *Ecological economics: The science and management of sustainability.* New York: Columbia University Press.

Cormack, W.F., 1997. Testing the sustainability of a stockless arable rotation on a fertile soil in Eastern England. In: Isart, J. and Llerena, J.J. (eds). *Resource use in organic farming*. Proceedings of the third ENOF workshop, Ancona, 5-6 June, 1997. Barcelona: The European Network for Scientific Research Coordination in Organic Farming (ENOF).

Council of Europe 1998. *The Draft European Landscape Convention*. Strassbourg: Council of Europe, 16p.

Dabbert, S., 1997. Support of organic farming as a policy instrument for resource conservation. In: Isart, J. and Llerena, J.J. (eds). *Resource use in organic farming*. Proceedings of the third ENOF workshop, Ancona, 5-6 June, 1997. Barcelona: The European Network for Scientific Research Coordination in Organic Farming (ENOF).

Daly, H.E. and Townsend, K,E., 1993. *Valuing the earth: Economics, ecology, ethics*. London: The MIT Press.

Daly, H.E., Cobb, J.B. and Cobb, C.W., 1990. *For the common good: Redirecting the economy towards community, the environment and a sustainable future*. London: Green Print.

De Bruin, H., 1997. *Dynamiek en duurzaamheid: Beschouwingen over bedrijfsstijlen, bestuur en beleid*. Wageningen: Wageningen Agricultural University.

De Bruin, R., Van Broekhuizen, R. and Van der Ploeg, J.D., 1997. *Ondernemen van onderop: Plattelandsvernieuwing in Gelderland*. Wageningen: Wageningen Agricultural University.

De Bruin, R.R., Oostindie, H. and Van der Ploeg, J.D., 1992. *Verbrede plattelandsontwikkeling in de praktijk*. Studierapporten van de Rijksplanologische Dienst no. 54. The Hague: VROM, 69pp.

De Groot, R.S. and Wagenaar Hummelinck, M.G., 1992. *Functions of nature: evaluation of nature in environmental planning, management and decision making*. Groningen: Wolters-Noordhoff, 315 p.

De Hen, P, and Van Leeuwen, A., 1997. Van wie is Nederland? De eigenaren van vier miljoen hectaren. *Elsevier*, 53 (2), pp: 12-17.

De Jong, T.M. 1978. *Milieudifferentiatie*. Technische Hoge school Delft, dissertatie. The Hague: Staatsuitgeverij, 312 pp.

De Putter, J., 1995. *The Greening of Europe's agricultural policy: The agri-environmental regulation of the MacSharry reform*. The Hague: Ministry of Agriculture, Nature Management and Fisheries, Agricultural Economics Research Institute LEI-DLO, 157p.

De Vries, W.M., 1994. *Pluri-activiteit in de Nederlandse landbouw*. Studies van landbouw en platteland nr 17. Wageningen: Circle for Rural European Studies, Wageningen Agricultural University.

De Wit, C.T., 1992. Resource use efficiency in agriculture. *Agriculture, Ecosystems & Environment*, 40, pp:125-151.

Defrancesco and M. Merlo 1996. Landscape values in farming and forestry environmental accounting: Area scale versus enterprise approach. In: S. Dabbert and Umstaetter, J., 1996. *Policies for landscape and nature conservation in Europe: an inventory to accompany the workshop on Landscape and nature conservation, held on 26th-29th September, 1996 at the University of Hohenheim*. Stuttgart: University of Hohenheim, 238 p.

Dekkers, Th.B.M. and Van der Werff, P.A., 1996. Biological processes in phosphate management. In: Van Ittersum, M.K., Venner, G.E.G.T., Van de Geijn, S.C. and Jetten, T.H. (eds) *Book of Abstracts, Fourth Congress of the European Society For Agronomy*, Veldhoven/Wageningen, 402p.

Devall, B. and Session, G., 1985. *Deep Ecology, living as if Nature Mattered*. Layton Smith (US): George Sessions, 267p.

DLG (Dienst Landelijk Gebied), 1997. *Nature Conservation on the Farm*. Utrecht: Edwards.

Dokuchayev, 1892. In: Kashtanov, 1992. *Nauchnoye naslediye V.V. Dokuchauev I yevo razvitye v sovremennon Inaschaftnom zemledelii* (The scientific heritage of V.V. Dokuchayev and its meaning for present agriculture). Keynote paper presented at the Scientific Conference of the Russian Academy of Agricultural Sciences, Vorohesh, Russia.

Dousek, J., 1995. Animal welfare and animal husbandry. *Veterinarstvi*, 45: 4, p181.

Draghetti, A., 1990. *Principi di fisiologia dell' azienda agraria*. Edagricole, Bolgna, 416p.

ECNC (European Centre for Nature Conservation), 1994. *Natural environment and sustainable development; Habitats, species and human society*. A network project of the European Centre for Nature Conservation. Tilburg: European Centre for Nature Conservation, 24p.

Edwards, 1984. Nature Conservation on the Farm. In: Green B.H., 1990. Agricultural intensification and the loss of habitat, species and amenity in British grasslands: a review of historical change and assessment of future prospects. Grass and Forage Science, vo. 45, pp:365-372.

Edwards-Jones, G. and Howells , O., 1997. An analysis of the absolute and relative sustainability of the crop protection activity in organic and conventional farming systems. In: Isart, J. and Llerena, J.J., (eds). *Resource use in organic farming*. Proceedings of the third ENOF workshop, Ancona, 5-6 June, 1997. Barcelona: The European Network for Scientific Research Coordination in Organic Farming (ENOF).

Eksesbo, I., 1992. Animal welfare. Ehtics and Welfare in meat animal husbandry. *Meat focus International*, 1; 6, pp:283-288.

El Titi, A., 1992. Integrated farming: an ecological farming approach in European agriculture. *Outlook Agriculture Oxon*, 21, 1, pp:33-39.

Ellenberg, H. 1988. *Vegetation ecology of Central Europe*. Cambridge: Cambridge University Press, 731pp.

Elzakker, B., Witte, R. and Mansvelt, J.D. van, 1992. Benefits of Diversity. New york:UNDP, 209p.

Estupinan, L.C., Isart, J., Vilata, F. and Isart, C., 1997. Improvement of human knowledge as resource for the setting up of ecological agricultural practices: the experience in Colombia. In: Isart, J. and Llerena, J.J. (eds). *Resource use in organic farming*. Proceedings of the third ENOF workshop, Ancona, 5-6 June, 1997. Barcelona: The European Network for Scientific Research Coordination in Organic Farming (ENOF).

Etienne, M., 1996. *Integrating livestock grazing into Mediterranean forest management as a fire prevention tool*. In: La foret paysanne dans l' espace rural. Versailles, INRA, SAD Etudes et Recherches no. 29.

Evans, H.E. and Ngau, P., 1991. Rural-urban relations, houdehold income diversification and agricultural productivity. *Development Change*, 22, pp:519-545.

Faeth, P., 1993. Evaluating agricultural policy and the sustainablity of production systems: An economic framework. *Journal of Soil and Water Conservation*, 48, pp:94-99.

Faeth, P., Repetto, R., Kroll, K., Dai, Q., and Helmers, G., 1991. *Paying the farm bill: U.S. agricultural policy and the transition to sustainable agriculture*. Washington, D.C.: World Resources Institute).

Fairbrother N., 1974. *The nature of landscape design*. London: Architectural press.

FAO, 1991. *The Den Bosch Declaration and Agenda for Action on Sustainable Agriculture and Rural Development (SARD)*. Main Document 1. Rome: FAO, 29p.

FAO, 1992. *FAO policies and actions. Stockholm 1972 - Rio 1992.* Rome: FAO, 88p.

Fleury, P. and Muller, S., 1995. The different components of bio-diversity in grasslands. Examples from the northern French Alps. *Acta Botanica Gallica*, 143, 4-5, pp:291-298.

Forman, R.T.T. and Gordon, M., 1986. *Landscape ecology.* New York: John Wiley, 619 pp.

FWAG (Farming and Wildlife Advisory Group), 1995. *The Whole Farm Conservation.* Plan for Elm Farm Organic Research Centre. Oxford: Berks & Oxon Farming and Wildlife Advisory Group Report, Reference No. ACS/BK4/15.05.95.

Gama, L.T., Matos, C.P., Matassino, D. (eds.) 1996. *Current situation of animal genetic resources in Portugal.* Wageningen: Wageningen Press.

Garcia 1992. Conserving the species-rich meadows of Europe. In: *Agriculture, Ecosystems and Environment*, 40: pp:219-232.

Gengenbach, H. and Limbacher, M, 1991. *Kooperation oder Konkurs?* Stuttgart: Verlag Freies Geistesleben, 170p.

George, S., 1985. *How the other half dies.* Harmordsworth, U.K.: Penguin, 352p.

Giorgis, S. 1995. *Rural landscapes in Europe: principles for creation and management.* Strasbourg: Council for Europe, 72p.

Godden, B. and Pnninckx, M., 1997. Mangement of farmyard manure composting is important to maintain sustainability in organic farming. In: Isart, J. and Llerena, J.J. (eds). *Resource use in organic farming.* Proceedings of the third ENOF workshop, Ancona, 5-6 June, 1997. Barcelona: The European Network for Scientific Research Coordination in Organic Farming (ENOF).

Goldhammer, J.G. and Jenkins, M.J., 1990. *Fire ecosystem dynamics: Mediterranean and Northern perspectives.* The Hague: SPB Academic Publishing, 199p.

Gore, A., 1992. *Earth in balance. Forging a New Common Purpose.* London: Earthscan, 407p.

Green B.H., 1990. Agricultural intensification and the loss of habitat, species and amenity in British grasslands: a review of historical change and assessment of future prospects. *Grass and Forage Science*, 45, pp:365-372.

Greenpeace, 1992. *Green fields, grey future; EC Agricultural policy at the crossroads.* Amsterdam: Greenpeace International, 98p.

Groh, T. and MacFadden, S.S.H., 1990. *Farms of Tomorrow, Community Supported Farms, Farm-supported communities.* Kimberton: Biodynamic Farming and Gardening Association, 169p.

Group of Bruges, 1996. *Agriculture, un tournant necessaire.* Paris: Editions de l' Aube, 98p.

Hagedorn, K., 1996. *Das Institutionenproblem in der agrarokonomischen Politikforschung.* Tubingen: J.C.B. Mohr, 551p.

Hagemeier, W., Tulp, I. and Groot, H., 1996. *Weidevogels in graslandgebieden van Nederland: trends en dichtheden.* Beek Ubbergen: SOVON, 85p.

Halberg, N., 1997. Farm level evaluation of resource use and environmental impact. In: Isart, J. and Llerena, J.J. (eds). *Resource use in organic farming.* Proceedings of the third ENOF workshop, Ancona, 5-6 June, 1997. Barcelona: The European Network for Scientific Research Coordination in Organic Farming (ENOF).

Halley, R.J. and Soffe, R.J., (Eds.) 1988. *Primrose McConnell's the agricultural notebook.- 18th edition.* Oxford: Blackwell Scientific Publications.

Hanley, N. (ed), 1991. *Farming and the countryside: An economic analysis of external costs and benefits.* Wallingford: CAB International.

Harris, J.M., 1996. World agricultural Futures: regional sustainability and ecological limits. *Ecological Economics Amsterdam*, 17, 2, pp95-115.

Harrisson, F., 1994. *The power of the Land.* New York: Springer, 223p.

Hassink, J., 1995. *Organic matter dynamics and N mineralisation in grassland soils.* Wageningen: Wageningen Agricultural University Diss., 250p.

Hayami, Y. and Ruttan, V.W., 1985. *Agricultural Development. An International Perspective.* Baltimore: John Hopkins University Press, 506p.

Haynes, L. 1994. *The sustainability of seed systems: a conceptual framework and a case study in Merida, Venezuela.* Wagenenigen: Wageningen Agricultural University, Department of Ecological Agriculture, MSc. Thesis, 119p.

Hemalata, R. *et al,* 1997. *Contours of social economic development policy issues.* New Delhi: Concept Publishing Company, 292p.

Hendriks, K. and Stroeken, F., 1993. *Verschil in verschijning een vergelijkende studie naar biologische en niet-biologische bedrijven.* Wageningen: Wageningen Agricultural University, Department Ecologische Landbouw, 109p.

Hendriks, K., Stobbelaar, D.J. and Van Mansvelt, J.D., 1997. Some criteria for landscape quality applied on an organic farm in Gelderland, the Netherlands. *Agriculture, Ecosystems & Environment,* 63 (2/3), pp. 185-201.

Hendriks, K., Stobbelaar, D.J. and Van Mansvelt, J.D., 1998. *Landschapswaarden van (biologische) landbouw: Een voorbeeld in West-Friesland (NH).* Wageningen: Wageningen Agricultural University, Department of Ecological Agriculture, 92pp.

Hiemstra, W., Reijntjes C. and Van der Werf, E., 1992. *Let Farmers Judge.* London: Intermediate Technology, 208p.

Hildyard, N., 1991. An open letter to Edouard Saouma, Director-general of the FAO. *Ecologist,* 21 (2).

Hobbelink, H. and Thompson, P.B., 1993. Biotechnology and the Future of World Agriculture. *Environmental Values,* (2/1), pp:83-84.

Hodge, I., 1991. The provision of public goods in the countryside: How should it be arranged? In: N. Hanley (ed) *Farming in the countryside: An economic analysis of external costs and benefits.* Wallingford: C.A.B. International.

Hoitink, H.A.J., 1989. *Disease suppressive properties of container media.* Wooster, Ohio: The Centre, pp:20-22.

Hoitink, H.A.J., Stone, A.G. and Han, D.Y., 1997. Suppression of plant diseases by composts. *Horticultural Science,* 32, (2), pp:184-187.

Hongxun, Y. 1982. *The classical gardens of China.* Van Nostrand Reinhold Company Inc.

Hoogendijk, W., 1993. *Economie onderste boven.* Utrecth: Jan van Arkel.

Howard, A. 1943. *An Agricultural Testament.* London: Oxford University Press, 253p.

Hueting, R., Bosch, P. and De Boer, B., 1992. *Methodology for the calculation of sustainable national income.* The Hague: CBS, 64pp.

Hund, K., 1994. *Een studie naar de inrichtingsmoglijkheden van de zone tussen Reve-/Abbertbos en Spijk/Bremerbergbos.* Enkhuizen, Amsterdam: Kees Hund T. & L. Architect BTN.

IEEP (Institute for European Environmental Policy), 1994. *The Nature of Farming: Low Intensity Farming.* London: Institute for European Environmental Policy.

IFOAM, 1992. Basic standards of organic agriculture. Tholey Theley, Germany: IFOAM General Secretariat, 32p.

IFOAM, 1996. *Basic Standards for Organic agriculture and Processing, and Guidelines for Coffee, Cocoa and Tea; evaluation of inputs.* Tholey Theley, Germany: IFOAM, 44p.

Isart, J. and Llerena, J.J. (eds), 1997. *Resource use in organic farming.* Proceedings of the third ENOF workshop, Ancona, 5-6 June, 1997. Barcelona: The European Network for Scientific Research Coordination in Organic Farming (ENOF).

Jenkis, N.R., 1987. Ecological changes resulting from a less intensive agriculture. In: Jenkis, N.R. and Bell, M. (eds). *Farm extensification: implications of EC Regulation 1760/87*, Merlewood: ITE.

Johnson *et al.*, 1951, cited in Odum, E.P., 1971.

Kabourakis, E., 1996. *Prototyping and dissemination of ecological olive production systems: A methodology for designing and a first step towards validation and dissemination of prototype ecological olive production systems (EOPS) in Crete.* Wageningen: Wageningen Agricultural University.

Kenmore, P., 1997. A perspective on IPM. *ILEIA-newsletter,* 13, pp:8-9.

Kenmore, P., Litsinger, J.A., Bandong, J.P., Santiago, A.C. and Salac, M.M., 1987, *Philippine rice farmers and insecticides: thirty years of growing dependency and new options for change.* Boulder: Westview Press.

Kenmore, P.E., 1991. *Indonesia's Integrated Pest Management - a model for Asia?* FAO Inter Country Programme for Integrated Pest Control in South and Southeast Asia, Manilla. Rome: FAO, 56p.

Kerkhove, G. 1994. *Sterk gemengd: een socio-economische analyse van agrarische bedrijvigheid in het Hageland en Pajottenland, Belgie.* Wageningen: Vlaams Agrarisch Centrum, Circle for Rural European Studies, Wageningen Agricultural University, 143p.

Kieft, H. and F. Verberne 1998. *Nieuwsgierig naar vernieuwing. Tussentijdse evaluatie van het beleidsthema "Plattelandsvernieuwing".* The Hague: VROM, LNV, IPO, 108p.

Kiley Worhtington, M., 1993. *Eco agriculture: food first farming: theory and practice.* London, Souvenir, 276p.

Kiss, J., Penksza, K., Toth, F. and Kadar, F., 1997. Evalaution of fields and margins in Nature production capcaity with special regard to plant protection. *Agriculture, Ecosystems & Environment*, 63 (2/3), pp: 227-233.

Klett, M., 1990. How can we liberate agriculture from its industrial prison? In: *Growing together.... Why should we bothter?* Hartfield, U.K.: International Biodynamic Initiative Group, 52p.

Kloen, H. and Vereijken, P., 1997. Testing and improving ecological nutrient management with pilot farmers. In: Isart, J. and Llerena, J.J. (eds). *Resource use in organic farming.* Proceedings of the third ENOF workshop, Ancona, 5-6 June, 1997. Barcelona: The European Network for Scientific Research Coordination in Organic Farming (ENOF).

Koepf, H., Schaumann, W. and Haccius, M., 1996. *Biological-dynamic agriculture.* Stuttgart: Ulmer, 376p.

Kouki, J., 1994. Biodiversity in the Fennoscandian forests: natural variation and its management. *Annales Zoologici Fennici*, 31, (1), pp:1-217.

Kovar, K. and Krasny, J., 1995. *Groundwater quality: remediation and protection.* Wallingford: IAHS Press, 500p.

Krekels, R., 1994. *Amfibieën en de ringslang terug in de Gelderse vallei.* Uitgave buro Natuurbalans i.s.m. Stichting Landschapsbeheer Gelderland.

Kuiper, J. 1997. Organic mixed farms in the landscape of a brook valley. How can a co-operative of organic mixed farms contribute to ecological and aesthetic qualities of a landscape? . *Agriculture, Ecosystems & Environment*, 63 (2/3), pp:121-133.

Kuiper, J. and Paans, L., 1990. *Landschapsstructuurplan voor de Zwijndrechtse Waard.* Wageningen: Wageningen Agricultural University, 90p.

Kuiper, J., 1998. *Ontwerpen in het rivierengebied: Landschapskwaliteit op verschillende schaalniveau's.*Wageningen: Wageningen Agricultural University, 128p.

Lai, R., 1989. In: Pimentel, D. and Hall C.W. (eds.) *Food and Natural Resources.* San Diego: Academic Press.

Lammerts, E.T., Hulscher, M., Jongerden, J., Haring, M., Hoogendoorn, J., Mansvelt, J.D.van, Ruivenkamp, G.T.P., 1998. *Naar een duurzame biologische plantenveredeling*. Driebergen: Louis Bolk Insitiuut, 62p.

Lampin, N.H., 1997. Organic livestock production and agricultural sustainability. In: J. Isart and Llerena, J.J. (eds). *Resource use in organic farming*. Proceedings of the third ENOF workshop, Ancona, 5-6 June, 1997. Barcelona: The European Network for Scientific Research Coordination in Organic Farming (ENOF).

Lampkin, N. and M. Measures (eds), 1995. *1995/96 Organic farm management handbook*. Aberystwyth: Elm: Universty of Wales, Elm Farm Research Centre.

Lampkin, N., 1990. *Organic farming*. Ipswich: Farming Press Books, 643p.

Lampkin, N.H. and S. Padel (eds), 1994. *The economics of organic farming: An international perspective*. Wallingford: CAB International.

Latacz-Lohmann, U., 1996. Mechanisms for the provision of public goods and in the countryside. In: Dabbert, S. and Umstaetter, J., 1996. *Policies for landscape and nature conservation in Europe: an inventory to accompany the workshop on Landscape and nature conservation, held on 26th-29th September, 1996 at the University of Hohenheim*. Stuttgart: University of Hohenheim, 238 pp.

Leake, A.R., 1997. An evaluation and comparison of energy resource usage in organic, integrated and conventional farming systtems: In: Isart, J. and Llerena, J.J. (eds). *Resource use in organic farming*. Proceedings of the third ENOF workshop, Ancona, 5-6 June, 1997. Barcelona: The European Network for Scientific Research Coordination in Organic Farming (ENOF).

Leloup, S.J.L.E., 1994. *Multiple use of a rangelands within agropastoral systems in southern Mali*. MSc. Thesis. Wageningen: Wageningen Agricultural University, 101 pp.

Lievegoed, B.C.J., 1979. *Phases*. London: Rudolf Steiner Press, 120p.

LNV, 1990. *Nature Policy Plan of the Netherlands*. The Hague: Ministry of Agriculture, Nature Management and Fisheries.

LNV, 1992. *Nota landschap: Regeringsbeslissing visie landschap*. The Hague: Ministry of Agriculture, Nature Management and Fisheries, 216 p.

LNV, 1993. *Notitie mest- en ammoniakbeleid derde fase*. The Hague: Ministry of Agriculture, Nature Management and Fisheries.

Lovelock, J.E., 1979. *Gaia, a new look at life on earth*. Oxford: Oxford University Press, 157p.

Lövenstein, H., Lantinga, E.A., Rabbingen, R., van Keulen, H., 1992. *Principles of Theoretical Production Ecology (TPE)*. Coursebook. Wageningen: Department of TPE, 111p.

Lundgren, F. and Friemel, J., 1994. *The nature of wealth: Discovering the physics within an economic system*. Kansas City, Missouri: The National Organization for Raw Materials, 106 p.

Lünzer and Kieffer, K.W., 1992. *Die Erde bewahren: Dimensionen einer umfassenden Oekologie*. Karlsruhe: Mueller, 392p.

Lynch, K., 1960. *The image of the city*. Cambridge: M.I.T. Press, 194 p.

Lynch, K., 1976. *Making Sense of a region*. Cambridge: M.I.T. Press, 238p.

MacNaeidhe, F.S., 1997. Nutrient budgeting for forage production using farmyard manure in an organic livestock system in Ireland. In: Isart, J. and Llerena, J.J. (eds). *Resource use in organic farming*. Proceedings of the third ENOF workshop, Ancona, 5-6 June, 1997. Barcelona: The European Network for Scientific Research Coordination in Organic Farming (ENOF).

Mäder, P., Pfiffner, L., Fleissbach, Wiemken, A. and Niggli, U., 1995. *Assessment of the soil microbial status under long-time low input (biological) and high inoput (conventional) agriculture.* In: Mäder, P., and Raupp, J., (eds) 1995. Effects of low and high external input agriculture and the soil microbial biomass and activities, in view of sustainable agriculture. Proceedings of the second meeting in Oberwill. Oberwill (CH): Research Institute for Organic Agriculture, 218p.

MAFF, 1996. *The Analysis of Agricultural Materials.* The Ministry of Agriculture, Food and Fisheries, B 427. London: HMSO.

Marino, D., Santucci, F.M., Zanoli, R. and Fiorani, S., 1997. Labour intensity in conventional and organic farming. In: Isart, J. and Llerena, J.J. (eds). *Resource use in organic farming.* Proceedings of the third ENOF workshop, Ancona, 5-6 June, 1997. Barcelona: The European Network for Scientific Research Coordination in Organic Farming (ENOF).

Masalles, R.M., Pino, J., and Sans, F.X., 1997. The role of population dynamics on the non-chemical weed control. In: Isart, J. and Llerena, J.J. (eds). *Resource use in organic farming.* Proceedings of the third ENOF workshop, Ancona, 5-6 June, 1997. Barcelona: The European Network for Scientific Research Coordination in Organic Farming (ENOF).

Maslow, A.H., 1968. *Motivation and personality.* New York: Harper and Row, 411 pp.

Mateu, F., Valle, M.A.N., Isart, J., Vilata, F. and Isart, J., 1997. Use of water as natural resource in Catalonia. Origin and quality. In: Isart, J. and Llerena, J.J (eds). *Resource use in organic farming.* Proceedings of the third ENOF workshop, Ancona, 5-6 June, 1997. Barcelona: The European Network for Scientific Research Coordination in Organic Farming (ENOF).

Mathes, M., 1997. *Ecological animal breeding in sustainable agriculture – An holistic appraoch to holistic criteria.* Braunschweig, FAO-conference on Sustainable Agriculture. In press.

Meadows, D.H. and Randers, J., 1992. *Beyond the Limits.* London: Earthscan, 300p.

Mearns, R., 1996. *When livestock are good for the environment: benefit sharing of environmental goods and services.* Brighton: University of Sussex, Institute of development Studies, 29p.

Meier Ploeger, A and Vogtmann, H., 1989. *A new approach to the determination of food quality.* Witzenhausen: Ekopan.

Michelakis, S.E., 1997. The insect pest management in organic olive and citrus groves of Crete, Greece, with the use of parasitoids, mass trapping and cultural methods. In: Isart, J. and Llerena, J.J. (eds). *Resource use in organic farming.* Proceedings of the third ENOF workshop, Ancona, 5-6 June, 1997. Barcelona: The European Network for Scientific Research Coordination in Organic Farming (ENOF).

Michelsen, J., 1997. Institutional preconditions for promoting conversion to organic agriculture. In: Isart, J. and Llerena, J.J. (eds). *Resource use in organic farming.* Proceedings of the third ENOF workshop, Ancona, 5-6 June, 1997. Barcelona: The European Network for Scientific Research Coordination in Organic Farming (ENOF).

Millar, P., 1997. *Livestock breeds in Great Brittain with a special regard to Scotland, and their place in breeding.*

Mol, D., 1993. Sustainable agriculture in the Netherlands. Wageningen: Wageningen Agricultural University, Department of Ecological Agriculture , 9p.

Molga, M., 1986. *Agro-meteorology.* Warsaw: PWRiL.

Mouchet, J. and Boudier, E., 1997. Sustainability audit of alternative dairy and beef cattle farms in western France. Methodological contribution and case study. In: Isart, J. and Llerena, J.J. (eds). *Resource use in organic farming*. Proceedings of the third ENOF workshop, Ancona, 5-6 June, 1997. Barcelona: The European Network for Scientific Research Coordination in Organic Farming (ENOF).

Murata, T. and Goh, K.M., 1997. Effects of cropping systems on soil organic matter in a pair of conventional and bio-dynamic mixed cropping farms in Canterburry, New Zealand. *Biology and fertility of soils*, 25, 4, pp:372-381.

Naess, A., 1975. Science between culture and counterculture. In : Dessauer, IC.I., Naess, A. and Reimer, E., (eds). *Science between culture and counterculture*. Congress paper. Nijmegen: Univeristy of Nijmegen (NL).

Naess, A., 1989. *Ecology, community and lifestyle: Outline of an ecosophy*. Cambridge: Cambridge University Press, 223 pp.

National Research Council, 1989. *Alternative Agriculture*. Washington: Washington National Academy Press, 448p.

Neher, D., 1992. Ecological sustainability in agricultural systems: definitions and measurements. *American Journal of Sustainable Agriculture*, 2 (3), pp:51-61.

NEPP3, 1998. *National Environmental Policy Plan 3: The Summary*. The Hague: VROM, 57p.

Neugebauer, B., Oldeman, R.A.A. and Valverde, P., 1996. Key principles in ecological silviculture. In: Oestergaard, T.V., 1996. *Fundamentals of organic agriculture*. 11th IFOAM International Scientific Conference, August 11-15, 1996, Copenhagen, Proceedings vol. 1. Tholey-Theley: IFOAM, 272pp.

Nijland, G.O. and Schouls, J., 1997. *The relationship between crop yield, nutrient uptake, nutrient surplus and nutrient application*. Wageningen: Wageningen Agricultural University, 151 pp.

Novakova, J., 1997. Agricultural impact on non-linear vegetation formation: species' richness – stand trophy relations. *Ekologia* 16(3), pp233-241.

O'Connor, R.J. and Shrubb, M., 1986. *Farming and birds*. Cambrige: Cambridge University Press.

Odum, E.P., 1971. *Fundamentals of Ecology*, third edition. Philadelphia, London, Toronto: W.B. Saunders Company.

OECD, 1994. *The contribution of amenities to rural development*. Paris: Organisation for Economic Co-operation and Development (OECD).

OECD, 1995. *Creating employment for rural development: New policy approaches*. Paris: Organisation for Economic co-operation and Development (OECD).

Oldeman, L.R., Hakkeling, R.T.A. and Sombroek, W.G., 1990. *World map of the status of human-induced soil degradation; World map, unexplanatory note and annex 5. International soil reference and information centre*. UN Environment programme in cooperation with the Winand Staring Centre, The International Society of Soil Science, Food and Agriculutral Organization of the UN and the International Institute for Aerospace Survey and Earth Science, 34pp.

Oldeman, R.A.A., 1990. *Forests: Elements of silvology*. Berlin: Springer Verlag, 624 pp.

Panayotou, Th., 1995. *Internalisation of environmental costs*. Draft prepared and presented at the UNEP/UNCTAD expert group meeting on internalisation of environmental costs, April 10-11, 1995. Cambridge: Harvard Institute for International Development.

Pauwels, F., Vanderhaegehe, S., Vandervoort, C., Van Hoorick, M.,, Verstraete, G. and Gulinck, H., 1996. Implementation of criteria for rapid assessment of landscape sustainability on organic and conventional farms in Belgium. In: Van Mansvelt, J.D., Stobbelaar, D.J. and De Graaf, J. (eds) 1996, *The landscape and nature production capacity of organic/sustainable types of agriculture.* Proceedings of the third plenary meeting of the EU Concerted Action AIR3-CT93-1210, pp23-31. Wageningen: Wageningen Agricultural University, Department Ecological Agriculture.

PEBLDS, 1995. *Pan-European biological and landscape diversity strategy.* Text transmitted to the UN-ECE Working Group of senior governmental officials environment for Europe. Strasbourg: Steering Committee for the Protection and Management of the Environment and Natural Habitats (CDPE) of the Council of Europe, 30pp.

Peeters, A., Lambert J., Lambert R., Janssens, F., 1993. *Diverse grassland in Belgium*: In Isart, J. and Llerena , J.J. (eds). *Resource use in organic farming.* Proceedings of the third ENOF workshop, Ancona, 5-6 June, 1997. Barcelona: The European Network for Scientific Research Coordination in Organic Farming (ENOF).

Pimentel, D. (ed.), 1993. *World Soil Erosion and Conservation*, Cambridge: Cambridge University Press, 349p.

Pimentel, D., Harvey, C., Resosudarmo, P., Sinclair, K., Kurz, D., McNair, M., Crist, S., Shpritz, L., Fitton, L., Saffouri, R., and Blair, R., 1995. Environmental and Economic Costs of Soil Erosion and Conservation Benefits. *Science,* 267, pp: 1117-1121.

Pohl, N.E.L., 1995. *Der kommende Stadtpark. Über urbane Grundbefindlichkeiten und die Einmischung der Natur.* Dissertation T.U. Delft. Delft: Delft Technical University.

Poldervaart, P., 1996, Direct marketing of regional products. *Schweizerische Milchzeitung,* 123, 2, p4.

Preuschen, G., 1985. *The farmer as a protector of the soil.* Graz: Leopold Stocker, 140p.

Preuschen, G., 1993. Operational instructions for ecological agriculture. *Alternative Konzepte,* 83, Karlsruhe, 174p.

Price, C., 1993. Applied landscape economics: A personal journey of discovery. In: A.C. Flynn (ed) Costing the country side. Special issue of: *Journal of Environmental planning and Management,* 36 (1), pp:51-65.

Rønningen, K., 1996. Policies and measures for the cultural landscape in Norway and Western Europe. In: Dabbert, S. and Umstaetter, J., 1996. *Policies for landscape and nature conservation in Europe: an inventory to accompany the workshop on Landscape and nature conservation, held on 26th-29th September, 1996 at the University of Hohenheim.* Stutgart: University of Hohenheim, 238 pp.

Rabbinge, R., Rossing, W.A.H. and Van der Werf, W., 1990. *The bridge function of production ecology in pest and disease management.* In Rabbinge, R., Goudriaan, F. and Van Keulen, H., (eds). *Theoretical Production Ecology: Reflections and Perspectives.* Wageningen: Pudoc, 301p.

Ratheiser, N., 1996. *Viehwirtschaft: aktuelle rechtliche Grundlagen, Tierzucht – Fütterung – Tierhaltung – Tierschutz.* Vienna: Bundesministerium für land- und Forstwirtschaft.

Reijntjes, C. et al, 1992. *Farming for the Future: An introduction to low external input and sustainable agriculture.* Leusden: ILEIA, 250p.

Rist, M., Boehncke, E. and Schneider, M., 1992. Species appropriate cattle rearing. *Alternative Konzepte 77.* Karlsruhe, 232p.

RIVM (Rijksinstituut voor Volksgezondheid en Milieuhygiene) 1991. *Nationale Milieuverkenning 2, 1990-2010.* Alphen aan den Rijn: Samson, H.D. Tjeenk Wilink B.V.

RIVM (Rijksinstituut voor Volksgezondheid en Milieuhygiene) 1993. *Nationale Milieuverkenning 3, 1993-2015.* Alphen aan den Rijn: Samson, H.D. Tjeenk Wilink B.V.

RIVM, IKC-Nederland, DLO-IBN, and DLO-SC, 1997. *Natuurverkenning 1997.* Alphen aan den Rijn: Samson H.D. Tjeenk Willink B.V.

Roep, D., Van der Ploeg, J.D. and Leeuwis, C., 1991. *Zicht opp duurzaamheid en kontinuiteit; bedrijfsstijlen in de Achterhoek.* Wageningen: Wageningen Agricultural University, 207p.

Rossi, R., Nota, D. and Fossi, F. 1997. Landscape and nature production capacity of organic types of agriculture in two Tuscan landscapes. *Agr. Ecosyst. & Environment,* 63, pp:159-173.

Ryszkowski, L., 1995, Managing ecosystem services in agricultural landscape. *Nature and Resources,* 31, (3).

Ryszkowski, L., and Kedziora, A., 1987. Impact of agricultural landscape structure on energy flow and water cycling. *Landscape Ecology,* (1), pp:85-94.

Ryszkowski, L., and Kedziora, A., 1995. Modification of the effect of global climate change by plant cover structure in an agricultural landscape. *Geographia Polonica,* (65), pp::5-34.

Sage, C. and Redclift, M., 1994. *Population, consumption and sustainable development.* In: Sage, C. and Redclift, M. Eds. Strategies for sustainable development: local agendas for the Southern Hemisphere. Chichester: John Wiley and Sons.

Salm, A., 1997. *Direct connections: farmer-consumer communication in a local food system.* MSc thesis. Wageningen: Wageningen Agricultural University, Department of Ecological Agriculture, 102p.

Sambraus, H.H., 1994. *Endangered livestock breeds: their history, use and preservation.* Stuttgart: Eugen Ulmer Verlag, 384p.

Sarantonio, M., 1991. *Methodologies for screening soil improving legumes.* Pennysylvania, USA: Rodale Institute.

Sarapatka, B. and Zilka, M., 1997. Soil fertility enhancement in the agrosystem using modern technologies of organic manure treatment. In: Isart, J. and Llerena, J.J. (eds). *Resource use in organic farming.* Proceedings of the third ENOF workshop, Ancona, 5-6 June, 1997. Barcelona: The European Network for Scientific Research Coordination in Organic Farming (ENOF).

SARD, 1991. *Elements for strategies and agenda for action (draft proposal).* Rome: FAO and the Ministry of Agriculture, Nature Management and Fisheries of the Netherlands, 43 p.

Sattler, F. and Von Wistinghausen, E., 1992. *Biodynamic Practice.* Stourbridge (UK): BDAA, 33p.

Savory, A., 1995. *Holistic resource management.* Washington, D.C.: Island Press, 564 p.

Schad, W., 1993. Kann man von Kranken und gesunden Landshaft sprechen: In: Van Mansvelt, J.D. and Vereijken, J.F.H.M. (eds.) *Proceedings of the try-out workshop on values assessment in environment and landscape research.* A meeting with the Goethean approach. Bilthoven: RIVM Rep. 607042001, 59p.

Schama, S., 1995. *Landscape and Memory.* London: Harper Collins, 652p.

Schiller, F., 1981. *Ueber die aesthetische Erziehung des Menschen.* Muenchen: Hanser, 245p.

Schmitz, H., 1993. *Houtwallen, heggen en singels.* Utrecht: Stichting Landelijk Overleg Natuur- en Landschap-beheer.

Schotman, A., 1988. *Tussen bos en houtwal: broedvogels in een twents cultuurlandschap.* Leersum: RIN-rapport 88/37.

Schotveld, E. and Kloen, H., 1996. Onkruidbeheersisng in een multifunctionele vruchtwisseling. Wageningen: AB-DLO, 31p.

Schraps, W.G. and Schrey, H.P., 1997. Valuable soils in Norhtrhine Westphalia –soil scientific criteria for generating a complete map for soil protection. *Zeitschrift fur Pflanzenernahrung und Bodenkunde,* 160, (4), pp:407-412.

Schultink, G., 1992. Evaluation of sustainable development alternatives: relevant concepts, resources assessment appraoches and comparative spatial indicators. *International Journal of Environmental Studies,* 41, pp:203-224.

Schumacher, E.F., 1973. *Small is beautifull.* London: Blond and Briggs, 138p.

Seel, M., 1991. *Eine Aesthetik der Natur.* Frankfurt am Mein: Suhrkamp, 388 p.

Sethuraman, S. and Ahmed, A., 1992. *Urbanisation, employment and the environment.* A world employment study. Geneva: International Labour Office.

Sheldrake, R., 1990. *The rebirth of nature: the greening of God.* London: Century, 215p.

Silva Pando, F.J. and Gonzales Hernandez, M.P., 1992. Agroforestry helps to prevent forests fires. *Agroforestry Today,* 4, (4), pp:7-8.

SKAL, 1997. *Skal normen als bedoeld in artikel 3, eerste lid sub d, van de statuten van de Stichting Skal.* Zwolle: Stichtinfg Skal, 50p.

Smeding, F., 1994. *Natuurwaarden van graslandpercelen en perceelsranden op twee alternatieve landbouwbedrijven.* Wageningen: Wageningen Agricultural University, Department Ecological Agriculture, 83p.

Smeding, F.W., 1995. *Protocol natuurplan.* Wageningen: Wageningen Agricultural University, Department Ecological Agriculture, 137p.

Sotherton, M.R., 1985. The distributiion and abundance of predatory Coleoptera overwintering in field boundaries. *Ann. Appl. Biol.,* 106, pp:17-21.

Soule, M.E. and Simberhoff, D., 1986. What do genetics and ecology tell us about the design of nature reserves? *Biological Conservation,* 35, pp:19-40.

Spinelli, R. and Baldini, S., 1993. The role of the draught horse in forestry operations. *Monti e Boschi,* 44, (6), pp:26-33.

Steiner, R., 1924. *Agriculture, a course of eight lectures.* London: Biodynamic Agriculture Association, 1974, 258p.

Stobbelaar, D.J. and Van Mansvelt, J.,D. (Eds.) 1994. *The landscape and nature production capacity of organic/sustainable types of agriculture.* Proceedings of the first pleanry meeting of the EU Concerted Action AIR3-CT93-1210, Wagenigen: Wageningen Agricultural University, Department of Ecological Agriculture, 194p.

Stopes, C., Measures, M., Smith C., Foster L., 1995. Hedgerow management in organic farming - impact on biodiversity. In: Isart, J. and Llerena, J.J.: Biodiversity and Land Use: The Role of Organic Farming. Proceedings of the first ENOF Workshop, Bonn, 8-9 December, 1995.

Stroeken, F., Hendriks, K., Kuiper, J., and Van Mansvelt, J.D., 1993. Veschil in verschijning. Een vergelijking van vier biologische en aangrenzende gangbare landbouwbedrijven. *Landschap,* 1993, 2, pp:33-45.

Struif Bontkes, T., 1998. *Modelling the dynamics of agricultural development: a process approach.* Wageningen: Wageningen Agricultural University, Diss., 232p. (in press).

Sundrum, A., 1993. Tierschutznormen in der ökologischen Nutztierhaltung und Möglichkeiten zu ihrer Kontrolle. *Deutsche Tierarztliche Wochenschrift,* 100, (2), pp:71-73.

Tape, N., 1992. *Joint FAO/WHO food standards programme.* Rome: FAO, 30p.

Tellarini, V., Caporali, F. and Di Lacovo, F., 1996. Some observations, not merely economic, concerning sustainable agriculture, as well as possible evaluation criteria and parameters. In: Van Mansvelt, J.D. and Stobbelaar, D.J., 1996. *The landscape and nature production capacity of organic/sustainable types of agriculture.* Proceedings of the third pleanry meeting of the EU Concerted Action AUR3-CT93-1210, Wagenigen: Wageningen Agricultural University, Department of Ecological Agriculture.

Temirbekova, S. and Mansvelt, J.D. van, 1998. *Theoretische und Praktische Möglichkeiten der Adaptation im biologischen Landbau.* Michneva: MOVIR, 75p (in press).

Terrasson, Daniel, Sophie Le Floch, 1997. *Landscape research priorities in France.* Paper presented on the WLO congress November 1997.

Thiele, H.U., 1997. Carabid *Beetles in Their Environments.* Berlin, Heidelberg, New York: Springer Verlag, 369p.

Thompson, K., Hogdson, J.G., Grime, J.P., Rorison, I.H., Band, S.R., and Spencer, R.E., 1993. Ellenberg numbers rivisited. *Phytocoenologica,* 23, pp:277-289.

Troeh, F.R. and Thompson, L.M., 1993. *Soils and Soil Fertility.* New York: Oxford University Press, New York.

Tucker, G.M., and Heath, M.F., 1994. *Birde in Europe: their conservation status.* Birldife Conservation Series No.3. Cambridge: Birdlife International.

Tyrrell, L. E., and Crow, T.R., 1994. Dynamics of dead wood in old grown hemlock-hardwood foersts of northern Wisconsin and northern Michigan. *Canadian Journal of Forestry Research,* 24, (8), pp:1672-1683.

UNCED, 1992. *Agenda 21 of the Earth Summit.* New York: United Nations, 124p.

UNCED, 1993. *The Earth Summit: The United Nations Conference on Environment and Development.* (S.P. Johnson). London: Graham T. Trotman, 532p.

UNDP, 1991. *Human development report.* New York: Oxford University Press, 202p.

Vahle, H.C., 1993. Die Idee der Kulturlandschaft. In: Van Mansvelt, J.D. and Vereijken, J.F.H.M. (eds). *A meeting with the Goethean approach. Proceedings of the try-out workshop: Values assessment in environment and landscape research.* Bilthoven: RIVM, report no. 607042001.

Van Bol, V., and Peeters, A., 1995. *Nature conservation in the framework of an Ecological Farming System.* In: Isart, J. and Llerena, J.J. *Biodiversity and Land Use: The Role of Organic Farming.* Proceedings of the first ENOF Workshop, Bonn, 8-9 December.

Van Broekhuizen, R. and Van der Ploeg, J.D., 1997. *Over de kwaliteit van plattelandsontwikkeling: Opstellen over doeleinden, sociaal-economische impact en mechanismen.* Studies van Landbouw en Platteland nr 24. Wageningen: Circle for Rural European Studies, Wageningen Agricultural University.

Van Buel, H., 1996. *Weidevogels binnen en buiten realtienota gebieden in Zeeland in 1995.* De Horst: LBL, 120p.

Van den Ham, A., Verstegen, J.A.A.M. and Greven, H.C., 1998. *Meer natuur op landbouwbedrijven: "Dus wij doen het niet goed?"* Wageningen: Landbouw-Economisch Institute (LEI-DLO) and Instituut voor Bos- en Natuuronderzoek (IBN-DLO).

Van der Lubbe, M.J., 1996. *Internalising ecological costs and benefits: A literature review.* Leusden: ETC Netherlands.

Van der Maarel, 1993. Relations between sociological-ecological species groups and Ellenberg indicator values. *Phytocoenologia,* 23, pp:343-362.

Van der Ploeg, J.D., and Ettema, M., (eds) 1990. *Tussen Bulk en Kwaliteit: Landbouw, Voedselproductie en Gezondheid.* Assen, Maastricht: Van Gorcum, 149p.

Van der Ploeg, J.D. and Roep, D., 1990. *Bedrijfsstijlen in de Zuidhollandse veenweidegebieden: Nieuwe perspectieven voor beleid en belangenbehartiging.* Wageningen, Haarlem, 98p.

Van der Ploeg, J.D., 1997. Groei vermoeit, zelfvertrouwen blijft. *Boerderij*, no.1 (30 September, 1997), pp: 6-14.

Van der Werff, P.A., Baars, T., and Oomen, G.J.M., 1995. Nutrient balances and measurement of nitrogen losses on mixed ecological farms on sandy soils in the Netherlands. *Biological Agriculture and Horticulture,* 11, 1/4, pp:41-50.

Van der Windt, H.J., 1995. *En dan: wat is natuur nog in dit land?* Amsterdam: BOOM, 336 p.

Van Dieren, W. 1995. *Taking nature into account: toward a sustainable national income: A report to the Club of Rome.* New York: Copernicus, 332p.

Van Elzakker, B., Witte, R. and Van Mansvelt, J.D., 1992. *Benefits of Diversity.* New York: UNDP, 209p.

Van Heudsen, W.R.M., Bruins, M., Hermens, E.M.P., Visers, J., 1994. *Ideeënboek beplantingen.* LD-mededeling nr. 202, Werkdocument Nr. 62, Wageningen: IKC-Natuurbeheer.

Van Mansvelt, J.D., 1981. *Alternative Landbouw: Landbouw doen, met hart en ziel, zonder je hoofd te verliezen.* Intree Rede. Wageningen: Wageningen Agricultural University, 40p.

Van Mansvelt, J.D., 1988. The role of lower-input technologies in the future. In: Whitbey, M. and Ollerenshaw, J. (eds). *Land use and the European environment.* London: Belhaven Press, 189p.

Van Mansvelt, J.D. and Verkley, F., 1991. *Society's steps toward sustainable agriculture.* UNESCO Paper SHS-91, Conf.802/14. Paris: UNESCO, 44p.

Van Mansvelt, J.D., and Mulder, J.A., 1993. European Features for Sustainable Development: a contribution to the dialogue. *Landscape and Urban Planning*, 27, pp:67-90.

Van Mansvelt, J.D. and Van Elzakker, B., 1994. *Policy making towards organic agriculture development.* In: Znaor, D., (ed). 1994: *The contribution of organic agriculture to sustainable rural development in central and Eastern Europe,* Ede: Avalon seminar for policymakers, Rudolec (Czech Republic), Oct. 1993, 222p.

Van Mansvelt, J.D. and Stobbelaar, J.D. and De Graaf, J. (eds.), 1995. *The Landscape and nature production capacity of organic/ sustainable types of agriculture.* Proceedings of the second plenary meeting of the EU-concerted action AIR3-CT93-1210. Wageningen: Wageningen Agricultural University, Department Ecological Agriculture, 247 p.

Van Mansvelt J.D., Stobbelaar, J.D. and De Graaf, J. (eds), 1996. *The landscape and nature production capacity of organic/sustainable types of agriculture.* Proceedings of the third plenary meeting of the EU Concerted Action Action AIR3-CT93-1210, Wageningen Agricultural University. Department of Ecological Agriculture, Wageningen, 129p.

Van Mansvelt J.D., Stobbelaar, J.D. and Hendriks, K., 1996. Landscape values of (organic) agriculture supported under regulation 2078/92. Consolidated report for EU. In: Kaule, G., and Morgan, M. (eds) *Regional guidelines to support sustainable landuse by EU agri-environmental programmes AIR 3 CT94-1296.* Stuttgart: University Stuttgart.

Van Mansvelt, J.D., 1997. An interdisciplinary approach to integrate a range of agro-landscape values as proposed by representatives of various disciplines. *Agriculture Ecosystems & Environment,* 63, pp: 233-250.

Van Mansvelt, J.D. and Stobbelaar, D.J., (eds) 1997. Landscape values in agriculture: strategies for the improvement of sustainable production. Special issues: *Agriculture, Ecosystems & Environment,* 63 (2/3). Amsterdam: Elsevier, 249 pp.

Van Mansvelt, J.D. and Van Laar, J.N., 1998. *The landscape and nature production capacity of organic/sustainable types of agriculture*. Progress report 4 of the Concerted Action AIR3-CT93-1210: Wageningen: Wageningen Agricultural University, Department of Ecological Agriculture.

Van Pelt, M.J.F., 1993. *Ecological sustainability and project appraisal: case studies in developing countries*. Avebury: Aldershot, 280p.

Van Vliet, I., 1998. *Zijn biologische producten gezonder dan gangbare?* Wageningen: Wageningen Agricultural University, Fonds Wetenschapswinkel, 176p.

Vereijken, J.F.H.M., 1995. *Identiteit, een uitdaging voor het andschapsbeheer.* Annual Report of the Louis Bolk Insitute. Driebergen: Louis Bolk Institute, pp:5-18.

Vereijken, J.F.H.M., T. van Gelder and Baars, T., 1997. Nature and landscape development on organic farms. *Agriculture, Ecosystems & Environment*, 63 (2/3), pp: 201-221.

Vereijken, P., 1986. Maintenance of soil fertility on bio-dynamic farms in Nagele, NL. In: Vogtman, H., Boehnke, E. and Fricke, I., (eds) 1986. *Oeko-landbau: eine weltweite notwendigkeit*. Karlsruhe: Muller, 289p.

Vereijken, P., 1992. A methodical way to more sustainable farming system. *Netherlands Journal of Agricultural Science,* 40, pp:209-223.

Vereijken, P., 1994. *Designing prototypes: Progress reports of research network on integrated and ecological arable farming systems for EU and associated countries (Concerted Action AIR 3-CT920755)*. Wageningen: DLO Research Institute for Agrobiology and Soil Fertility.

Vereijken, P. and Kloen, H., 1994. *Innovative research with Ecological Pilot Farmers*. In: Struik P.C., *et al.* (eds). Plant production on the threshold of a new century. Dordrecht and Boston: Kluwer, pp: 37-56.

Vereijken, P., 1995. *Designing and testing prototypes: Progress report 2 of research network on integrated and ecological arable farming systems for EU and associated countries*. Wageningen: DLO Research Institute for Agrobiology and Soil Fertility.

Vereijken, P., 1996a. *Progress Report 3: testing and improving prototypes*. Research Network for EU and Associated Countries on Integrated and Ecological Arable Farming Systems. Wageningen: DLO Research Institute for Agrobiology and Soil Fertility.

Vereijken, P., 1996b. *Prototyping integrated and ecological arable farming systems (I/EAFS) within an EU network*. Paper presented at Second European Symposium on Rural and Farming Systems. Wageningen: DLO Research Institute for Agrobiology and Soil Fertility.

Vereijken, P., 1998. *Improving and disseminating prototypes*. Progress report no. 4 of the research network on integrated and ecological arable farming systems for EU and associated countries. Wageningen: AB-DLO.

Verhoog, H., 1980. *Science and the social responsibility of natural scientists*. Diss. Leiden: Leiden University, 225p.

Vogtman, H. (ed.), 1985. *Oekologischer Landbau, Landwirtschaft mit Zukunft*. Stuttgart: Pro Nature Verlag, 159pp.

Volker, K., 1997. Local commitment for sustainable rural landscape development. *Agriculture, Ecosystems & Environment*, 63 (2/3), pp107-121.

Vollenbroek, J., 1994. Inventory of emissions into surface water in the Danube river basin: The role of agriculture. In: D. Zanor (ed.) *The contribution of organic agriculture to sustainable rural development in Central and Eastern Europe*. Ede: Avalon.

Von Fragstein, P., 1996. Nutrient management in organic farming. In: Ostergaard, T.V. (ed.) *Fundamentals of Organic Agriculture*. Proceedings of the 11th FOAM International Scientific Conference, Copenhagen, IFOAM, 1, pp:62-72.

Von Mallinckrodt, F., 1991. *More efficient agriculture without losing the physical and moral foundation.* New York: UNDP, 83p.

Von Borell, E., 1996. *Current situation on welfare legislation and research within the European Union.* Halle: Insititute of Animal Breeding and Husbandry.

Von Weiszäcker, E., Lovins, A.B. and Lovins, L.H., 1997. *Factor Four: doubling wealth-halving resource use: the new report to the Club of Rome.* London: Earthscan, 322p.

Vos, W. and Stortelder, A. 1992. *Vanishing Tuscan landscapes: Landscape ecology of a submediterranean montane area (Solano Basin, Tuscany, Italy).* Wageningen: Pudco, 404pp.

Vos, W. and Fresco, L.O., 1994. Can agricultural practices contribute to functional landscapes in Europe? In: D.J. Stobbelaar and J.D. van Mansvelt (eds*) Proceedings of the first plenary meeting of the EU-Concerted action.* Wageningen: Wageningen Agricultural University, Department of Ecological Agriculture.

Vroom, J.M., 1986. The perception of dimensions of space and levels of infrastructure and its application in landscape planning. *Landscape Planning*, 1986, 12, pp:337-352.

Wallin, K., 1985. Spatial and temporal distribution of some abundant carabid beetles (Coleoptera:Carabidae) in cereal fields and adjacent habitats. *Pedobiologia*, 28 pp:9-34.

Wander, M.M. and Traina, S.J., 1996. Organic matter fractions from organically and conventionally managed soils, II, characterisation of composition. *Soil Science Society of America Journal*, 60, (4), pp:1087-1094.

Wascher, D.M., 1996. Policy action to control environmental impacts from agriculture. In: Dabbert, S. and Umstaetter, J., 1996. *Policies for landscape and nature conservation in Europe: an inventory to accompany the workshop on Landscape and nature conservation, held on 26th-29th September, 1996 at the University of Hohenheim.* Stuttgart: University of Hohenheim, 238 pp.

Waterson, J., 1994. The draught horse in the UK forestry. *Quarterly Journal of Forestry*, 88, 4, pp:309-313.

Watson, C.A., Stopes, C. and Philipps, L., 1997. Nitrogen cycling in organically managed crop ratations: Importance of rotation design. In: Isart, J. and Llerena, J. (eds). *Resource use in organic farming.* Proceedings of the third ENOF workshop, Ancona, 5-6 June, 1997. Barcelona: The European Network for Scientific Research Coordination in Organic Farming (ENOF).

Werkgroep De Zeeuw, 1998: *Naar een Aartse Landbouw. Plattelandsontwikkeling en duurzame landbouw in een tijd van globalisering.* Amsterdam: IMSA, 60p.

Wilson, J., 1992, The BTO Birds and Organic Farming Project One Year On. *BTO News*, 185 pp:10-12.

Wiskerke, J.S.C., 1997. *Zeeuwse akkerbouw tussen verandering en continuiteit: Een sociologische studie naar diversiteit in landbouwbeoefening, technologieontwikkeling en plattelandsvernieuwing.* Wageningen: Wageningen Agricultural University.

Witte, R., Van Elzakker, B. and Van Mansvelt, J.D., 1993. *Rice and the environment: Environmental impact of rice production, policy review and options for sustainable rice development in Thailand and the Philippines.* Geneva: UNCTAD, 113p.

WLO 1998: *A research strategy for the next decade.* Amsterdam: The Dutch association for Landscape Ecology (WLO), 51p.

Worldwatch Institute, 1988. State of the World. Worldwatch Institute, Washington.

Yi-Fu Tuan, 1972. Discrepancies between environmental attitude and behaviour: examples from Europe and China. In: *Man space and environment.* London: Ed. English and Mayfield.

Younie, D. and Baars, T.1997. Resource use in organic grassland. The central bank and the art gallery of organic farming. In: Isart, J. and Llerena, J.J. (eds). *Resource use in organic farming*. Proceedings of the third ENOF workshop, Ancona, 5-6 June, 1997. Barcelona: The European Network for Scientific Research Coordination in Organic Farming (ENOF).

Znaor, D., 1996. *Ekološka poljoprivreda: poljoprivreda sutrašnjice*. Nakladni zavod "Globus", Zagreb, 469p.

ANNEX 1   PARTICIPANTS

The overall co-ordination, including the organisation of the plenary meetings and the writing and editing of the proceedings of the plenary meetings, was at the responsibility of J.D. van Mansvelt and D.J. Stobbelaar. The subgroup meetings are organised with the support of J. de Graaff and later J.N. van Laar, and the important local expertise of the participants whose countries are visited. The other participants joined the discussions in the meetings and thus contributed to the yearly reports and/or they have written a paper on their presentations at the meetings. The participants have been working towards an integration of their complementary methods where ever possible, or at least towards a clear identification of the incompatibility of their approaches.

The following scientists participated in the concerted action AIR3-CT93-1210 "The Landscape and Nature production Capacity of Organic/Sustainable Types of Agriculture":

Participants from European Union countries

Belgium

Marc Antrop
Department of Geography
University of Ghent
Ghent

Frans Pauwels
Catholic University
Leuven

Denmark

Katrine Højring
Danish Forest and
Landscape Research
Institute
Hørsholm

Germany

Michael Beismann
Goetheanum
Dornach

Sylvia Hermann
Department of Landscape planning
and Ecology
University of Stuttgart
Stuttgart

Thomas van Elsen
State University Kassel
Landscape Ecology and Nature
protection
Witzenhausen

Elisabeth Osinski
Department of Landscape planning
and Ecology
University of Stuttgart
Stuttgart

Giselher Kaule
Department of Landscape planning
and Ecology
University of Stuttgart

Great Britain

Margaret Colquhoun
Life Science Trust
Humbie, East Lothian
Scotland

Lynda Hepburn
Life Science Trust
Humbie, East Lothian
Scotland

Ulrich Loening
Centre for Human Ecology
University of Edinburgh
Edinburgh
Scotland

Ireland

Noel Culleton
Teagasc Research Centre
Co. Wexfort

Finnain MacNaeidhe
Teagasc Research Centre
Co. Wexford

France

Marc Benoit
INRA
Mirecourt

Waltroud Koerner
INRA
Mirecourt

Netherlands

Karina Hendriks
Department of Ecological Agriculture
Wageningen Agricultural University
Wageningen

Juliette Kuiper
Department of physical planning
and rural development
Wageningen Agricultural University
Wageningen

Jan Diek van Mansvelt
Department of Ecological Agriculture
Wageningen Agricultural University
Wageningen

Derk Jan Stobbelaar
Department of Ecological Agriculture
Wageningen Agricultural University
Wageningen

Hans Vereijken
Louis Bolk Institute
Department for biological and organic
agricultural research

Driebergen

Kees Volker
The Winand Staring Center (SC-DLO)
Wageningen

Lilian Hermens
IKC Nature Management
Wageningen

Frank Stroeken
Utrecht
The Winand Staring Center (SC-DLO)

Henk Doing
Department of Vegetation Ecology, Plant
Ecology and Weed Sciences
Wageningen Agricultural University
Wageningen

Italy

Roberto Rossi
Tuscany Region
Department of Agriculture and
Forestry
Florence

Vittorio Tellarini
Department of Agricultural
Economics
University of Pisa
Pisa

Maria Andreoli
Department of Economica
Aziendale
University of Pisa
Pisa

Spain

Gaston Remmers
Department of Sociology and Rural
Development
University of Cordoba
Cordoba

Juan Gasto
University of Cordboba
Cordoba

Miguel A. Herrera
University of Cordoba
Cordoba

Portugal

Ana Maria Firmino
New University of Lisbon
Faculty of Social and Human Sciences
Lisbon

Greece

Emmanouil Kabourakis
Cretan Agri-environmental group
Moires
Greece

Participants from countries outside the European Union

Estonia

Ülo Mander
Department of Geography
University of Tartu

Hungary

Jószef Kiss
Gödöllö University of Agricultura sciences
Department of plant protection
Gödöllö

Switzerland

Andreas Bosshard
Institute of Geo-botany ETH
Zürich

Max and Rosmarie Eichenberger
Nature Conservation and Agriculture, Rodersdorf

Norway

Morten Clemetsen
Aurland

Gary Fry
Norwegian Institute for Natur Research
Centre

ANNEX 2   CHECKLIST'S COMPLIANCE WITH OTHER STANDARDS FOR SUSTAINABLE AND ORGANIC AGRICULTURE

D. Znaor

In this annex we present an attempt to check the conformity of the standards for the development of sustainable landscapes with some other standards for organic and sustainable agriculture:

1. IFOAM = (International Federation of Organic Agriculture Movements): Basic Standards for Organic Agriculture and Processing and Guidelines for Coffee, Cocoa and Tea; Evaluation of Inputs decided by the IFOAM General Assembly at Copenhagen/Denmark, August 1996.
2. EU = Official Journal of the European Communities: Council Regulation (EEC) No. 2092/91 of June 24 1991 on organic production of agricultural products and indications referring thereto on agricultural products and foodstuffs.
3. Vereijken, Kabourakis refer to the standards developed within the research network on integrated and ecological arable farming systems for EU and associated countries (Concerted Action AIR 3-CT920755), of which some are presented in the corresponding chapters in this book)

How to read this annex:
Following abbreviations are used and derived from IFOAM (1996):

GPri	IFOAM Standards: General principles.
PAim	IFOAM Standards: The Principle Aims of Organic Agriculture and Processing.
PReq	IFOAM Standards: The Principle Requirements of Organic Agriculture and Processing.
MRreq	IFOAM Standards: Minimum Requirements.
Inputs	IFOAM Guidelines on Evaluation of Inputs to Organic Agriculture.
CCT	IFOAM Guidelines for Coffee, Cocoa and Tea.
Annex I	Refers to the annex of the principles of Organic Production at Farm Level, EU Regulation.

Checklist nrs Refer to Chapter 3 of this report.

The numbers following an abbreviation in the "IFOAM" and "EU" column refer to the chapter/article of the original document. However, the numbers in the column "Vereijken, Kabourakis" do not refer to the chapters of the original document, but correspond to the checklist order as presented in this report.
Note that the IFOAM General Principles (GPri) and Principle Aims (PAim), Guidelines for Coffee, Cocoa and Tea (CCT), as well as Annex I of the EU Regulation reflect a

more spirit-like conformity with the proposed landscape standards. The IFOAM Minimum Requirements (MReq) and the Principle Requirements (PReq), as well as Guidelines on Evaluation of Inputs (Inputs) are more specific and qualitatively, conform with the proposed landscape standards. However, to what degree MReq and PReq comply with the proposed landscape standards is pretty arbitrary and the authors do not pretend to be sure about the full compliance among some particular criteria.

The standards developed by Vereijken and Kabourakis are more specific than those of IFOAM and EU. Their standards are an attempt to translate the above mentioned qualitative standards into quantitative standards and give clear numerical values (desired ranges) for some of the standards.

STANDARDS for the development of sustainable rural landscapes	Conformity with standards for organic/ sustainable agriculture		
	IFOAM	EU	Vereijken, Kabourakis
1. ENVIRONMENT (A-BIOTIC)			
1.1 Main criterion: Clean environment			
1.1.1 FERTILE AND RESILIENT SOIL	PAim, Gpri 4.3 & 4.6, MReq 4.7.2 & 4.7.5, CCT	Annex I/ 2	
PARAMETERS:			
1. *Manure quality (C/N ratio)*	PReq, MReq 4.7.4		
2. *Stocking rate matching carrying capacity*	MReq 4.7.2, CCT	Annex I/ 2a	
3. *anti-erosive belts and contour ploughing*			see checklist 1.1.1/ 3
4. *soil cover (winter or off-season)*	GPri 4.6.		see checklist 1.1.1/ 4
5. *crop-rotation / crop mixture*	GPri 4.2. & 4.9, CCT		see check list 1.1.1/ 5
6. *soil structure and organic matter content*	GPri 4.2. & 4.9		see check list 1.1.1/ 6
1.1.2 WATER QUALITY	PAim, GPri 4.2 & 4.6 & 4.9, PReq, MReq 4.7.2 & 4.7.5, CCT		
PARAMETERS:			
1. *cattle units/ha (all the farms' ha's)*	PReq, MReq 4.7.4		
2. *level and time of manuring (quantity/ha/yr)*	PAim, GPri 4.1, 4.3, 4.9, MReq 4.3.3. & 4.3.5		
3. *waste-water treatment*			see check list 1.1.2/ 4
4. *minerals and additives bookkeeping*	PAim (closed cycles), CCT		see check list 1.1.2/ 5
5. *other potential pollutants bookkeeping*	PAim, GPri 4.3, 4.7, MReq 4.7.3		see check list 1.1.2/ 6
6. *water use and management*			
1.1.3 AIR QUALITY			
PARAMETERS:			
1. *ammonia emissions*			
2. *other emissions*			see check list 1.1.3/ 2
3. *wind-shelter belts*			see check list 1.1.2/ 5
1.1.4 WILD FIRE CONTROL	MReq 4.7.1		

STANDARDS for the development of sustainable rural landscapes	Conformity with standards for organic/ sustainable agriculture		
	IFOAM	EU	Vereijken, Kabourakis
1. ENVIRONMENT (A-BIOTIC)			
1.2. Main criterion: food and fibre sufficiency and quality	PAim		
1.2.1 NATIONALLY SUFFICIENT AND REGIONALLY SUSTAINABLE LEVELS OF PRODUCTION	GPri 4.9		see checklist 2.1.2/ 1
PARAMETERS: 1. Minimal nutrient requirements per capita 2. Required area for sustainable agriculture 3. Level of integration of land for food production and land for nature production			
1.2.2 Good food and fibre quality to match sufficient quantit.	PAim		
PARAMETERS: 1. Self-balance in physiology of human organ. 2. Good sensorial and nutritional qualities 3. Regionally specific quality	GPri, MReq 7.1 & 7.2, Inputs 4		
.3. Main criterion: regional carrying capacity	PAim		

STANDARDS for the development of sustainable rural landscapes	Conformity with standards for organic/ sustainable agriculture		
	IFOAM	EU	Vereijken, Kabourakis
1. ENVIRONMENT (A-BIOTIC)			
1.4. Main criterion: economic and efficient use of resources			see checklist 1.4/ 1
PARAMETERS: *1. Resource efficient energy management*	PAim (re-use, recycle), PReq		
2. Minim. required input of non-renew. energy	PAim (re-use, recycle), Preq		
3. Dependence on non-renew. energy sources			
4. Net yield from external non-renewab. inputs			
1.5. Main criterion: sustainable, site-adapted and regionally specific production systems	PAim, GPri 4.1	Annex I/ 2a	
PARAMETERS: *1. Locally adapted farm management*	GPri 4.1 & 4.9, PReq,		
2. Cultivation of local crop and animal species	GPri 4.1 & 4.9		
3. Production of regionally speciality products	GPri 4.8		

STANDARDS for the development of sustainable rural landscapes	Conformity with standards for organic/ sustainable agriculture		
	IFOAM	EU	Vereijken, Kabourakis
2. ECOLOGY (BIO-SPHERE)			
2.1. Main criterion: bio-diversity	PReq		
2.1.1 FLORA AND FAUNA SPECIES' DIVERSITY	PAim (maintain gen. diver.) PReq, GPri 4.1 & 4.9		
PARAMETERS: 1. *Species diversity per bio-type and bio-tope* 2. *Targeted Plant Species Diversity (TPSD), Target Trees Index (TTI) and Target Shrubs Index (TSI)*			see checklist 2.1.1/ 2
3. *Plant Species Diversity (PSD) and Plant Species Distributions (PSDN)*			see checklist 2.1.1/ 3
2.1.2 BIO-TOPE DIVERSITY	PReq		
PARAMETERS: 1. *Minimum standard per bio-topes per farm type*			see checklist 2.1.2/ 1
2.1.3 ECO-SYSTEMS' DIVERSITY	GPri 4.9		
PARAMETERS: 1. *Minimum standard for types, numbers, and size of ecosystems per landsc. and region* 2. *Multifunctional landscape management* 3. *Regional specification on presence (quality) and abundance (quantity)*			

STANDARDS for the development of sustainable rural landscapes	Conformity with standards for organic/ sustainable agriculture		
	IFOAM	EU	Vereijken, Kabourakis
2. ECOLOGY (BIO-SPHERE)			
2.2. Main criterion: ecological coherence	PAim, PReq, GPri 4.9, Inputs 3, GPri 4.9		
2.2.1 VERTICAL COHERENCE: ONSITE	GPri 4.9		
PARAMETERS: 1. Site specific indicator species 2. Site specific habitats and ecosystems			
2.2.2 HORIZONTAL COHERENCE: IN THE LANDSCAPE	GPri 4.9		
PARAMETERS: 1. Species coherence 2. Habitat and eco-system coherence			
2.2.3 CYCLICAL COHERENCE: IN TIME			
PARAMETERS: 1. Full lifecycles of species and systems 2. Seasons compliant management: availability of nectar for 'flower-insects' 3. Seasons compliant management: timely differentiated hedge and woodland mngm. 4. Seasons compliant management: timely management of water-bodies 5. Seasons compliant management: timely management of permanent pastures			

STANDARDS for the development of sustainable rural landscapes	Conformity with standards for organic/ sustainable agriculture		
	IFOAM	EU	Vereijken, Kabourakis
2. ECOLOGY (BIO-SPHERE)			
2.3. Main criterion: eco-regulation	PReq, GPri & MReq 4.4	Annex I/ 3	
PARAMETERS: *1. Degree of pest and disease occurrence* *2. Pest predator presence*	GPri 4.4 GPri 4.9		
2.4. Main criterion: animal welfare condit.	PAim, PReq, GP 5.1-5.7 GPri 5.1, 5.2, 5.3		
PARAMETERS: *1. Space for natural behaviour* *2. Shelter against adverse weather* *3. Preventive health care*	MReq 5.1.1, 5.1.2, 5.1.3 MReq 5.1.1, 5.1.2, 5.1.3 GPri 5.4, 5.6, MReq 4.5.1, 5.4.1, 5.4.2, 5.4.7, 5.6.3		
3. ECONOMY			
3.1. Main criterion: good farming should pay-off	PAim		see checklist 3.1/ 1
PARAMETERS: *1. Total net farm income* *2. Total farm family income* *3. Return on labour* *4. Farm's market orientation* *5.Financial autonomy*			

STANDARDS for the development of sustainable rural landscapes	Conformity with standards for organic/ sustainable agriculture			
	IFOAM	EU	Vereijken, Kabourakis	
3. ECONOMY				
3.2. Main criterion: greening the economy	PAim, GPri 8.2		see checklist 1.4/ 1	
PARAMETERS:				
1. Technical autonomy				
2. Dependence on non-renewable inputs				
3. Share of re-used on-farm production value in total costs	PReq			
4. Share of non-renewable inputs in total costs				
5. The costs-benefits ratio of investments in landscape, environment and nature				
3.3. Main criterion: regional autonomy	PAim			
PARAMETERS:				
1. Transport	GPri 8.2			
2. Resource efficiency and regional labour possibilities	GPri 8.2			
3. Swaps from single community support to management system's support				
4. Translation of commodities under main criterion 3.1 and 3.2 to regional level	GPri 9.1			

STANDARDS for the development of sustainable rural landscapes	Conformity with standards for organic/ sustainable agriculture		
	IFOAM	EU	Vereijken, Kabourakis
4. SOCIOLOGY			
4.1. Main criterion: well-being in the area	PAim, GPri 8.1		
PARAMETERS: 1. Options for farmers' succession 2. Financial income 3. Welfare services in the region			
4.2. Main criterion: permanent education of farmers		PAim	
PARAMETERS: 1.Farmer's level of education 2. Farmer's participation in sust. agric. and landsc. relevant study circles and training			
4.3. Main criterion: access to participation			
4.3.1 FARMERS' INVOLVEMENT IN ACTIV. OUTSIDE THEIR FARMS			
PARAMETERS: 1. Membership to farmer organisat. and groups 2. Working in the region 3. Involvement in organising outlets 4. Co-operation with NGOs 5. Membership of regional councils 6. Access to professional expertise and support programme 7. Access to participate in dissemination progr.			

STANDARDS for the development of sustainable rural landscapes	Conformity with standards for organic/ sustainable agriculture		
	IFOAM	EU	Vereijken, Kabourakis
4. SOCIOLOGY			
4.3. Main criterion: access to participation			
4.3.2 OUTSIDERS' INVOLVEMENT IN FARM ACTIVITIES			
PARAMETERS: 1. Access to participate in landscape managm. 2. Professn.. and layman excursions to the farm 3. Community supported/shared agriculture 4. Financial commitment to landscape program. 5. Access, given to farmers, to buy/rent and manage landscape in a sustainable, ecolog. and socially sound way			
4.4. Main criterion: accessibility of the landscape			
PARAMETERS: 1. Excursions to the farm 2. Right of ways 3. Tracking roads			

STANDARDS for the development of sustainable rural landscapes	Conformity with standards for organic/ sustainable agriculture		
	IFOAM	EU	Vereijken, Kabourakis
5. PSYCHOLOGY			
5.1. Main criterion: compliance to natural environment			
PARAMETERS: 1. *Clear presence and cultivation (conservat.) of the region's special natural features like water-bodies of all sorts, slopes, peaks, marshes, dunes and cliffs.*			
5.2. Main criterion: good use of the land-scapes's potential utility			
PARAMETERS: 1. *Rationality of the -sustainable- land-use and the way it shows* 2. *Percentage of sustainable areas in proportion to the whole landscape and those managed in unsustainable ways* 3. *Possibilities for activities other than food and fibre production, on their feasible locations and their appropriate intensity of actual use*			

STANDARDS for the development of sustainable rural landscapes	Conformity with standards for organic/ sustainable agriculture		
	IFOAM	EU	Vereijken, Kabourakis
5. PSYCHOLOGY			
5.3. Main criterion: presence of naturalness PARAMETERS: 1. Indications that the landscape has developed in a sufficiently natural way 2. Dominance of natural elements, lines, patterns, materials, over artificial ones 3. Presence of natural, non-productive sites and old trees			
5.4. Main criterion: a rich and fair offer of sensory qualities PARAMETERS: 1. Smells 2. Sounds 3. Visual perceptions 4. Spatial perceptions			
5.5. Main criterion: experience of unity PARAMETERS: 1. Order 2. Completeness 3. Wholeness 4. Spaciousness			

STANDARDS for the development of sustainable rural landscapes	Conformity with standards for organic/ sustainable agriculture		
	IFOAM	EU	Vereijken, Kabourakis
5. PSYCHOLOGY			
5.6. Main criterion: experienced historicity			
PARAMETERS: 1. *Historic elements of art and crafts* 2. *Historic landscapes patterns*			
5.7. Main criterion: presence of cyclical development			
PARAMETERS: 1. *Developmental phases of natural elements* 2. *Landscape maintenance cycles* 3. *Succession of landscape bio-topes* 4. *Decomposition*			
5.8. Main criterion: careful management of the landscape			
PARAMETERS: 1. *Farm succession*			

STANDARDS for the development of sustainable rural landscapes	Conformity with standards for organic/ sustainable agriculture		
	IFOAM	EU	Vereijken, Kabourakis
6. PSYSIOGNOMY AND CULTURAL GEOGRAPHY			
6.1. Main criterion: diversity of landscape components PARAMETERS: 2. *Diversity of landscape types per country* 3. *Diversity of landscape units (bio-topes) per landscape type* 4. *Diversity of elements (crops and planting) per landscape unit* 5. *Diversity of species per bio-tope*			
6.2. Main criterion: coherence among landscape elements PARAMETERS: 1. *Hydrology* 2. *Infrastructure* 3. *Farming* 4. *Ecology*			
6.3. Main criterion: continuity of land use and spatial arrangement PARAMETERS: 1. *Cultural history* 2. *Duration and continuity of land use and spatial arrangement* 3. *Presumed future sustainability*			